Lifeline Engineering Systems

Jie Li · Wei Liu

Lifeline Engineering Systems

Network Reliability Analysis and Aseismic Design

Springer

Jie Li
College of Civil Engineering
Tongji University
Shanghai, China

Wei Liu
College of Civil Engineering
Tongji University
Shanghai, China

ISBN 978-981-15-9103-7 ISBN 978-981-15-9101-3 (eBook)
https://doi.org/10.1007/978-981-15-9101-3

Jointly published with Shanghai Scientific and Technical Publishers
The print edition is not for sale in China Mainland. Customers from China Mainland please order the print book from: Shanghai Scientific and Technical Publishers.

This Springer imprint is published by the registered company Springer Nature Singapore Pte Ltd.
The registered company address is: 152 Beach Road, #21-01/04 Gateway East, Singapore 189721, Singapore

Preface

One of the important signs showing the growth and prosperity of civil engineering in the twentieth century is the development and popularization of large-scale engineering networks. Regional engineering network systems such as electric power networks, urban water supply networks, gas supply networks, large-scale transportation networks, ..., all of them have a profound impact on human life and social progress. Because of their paramount importance, these engineering network systems are collectively called lifeline engineering systems or infrastructure systems.

However, in the early period of their development, the design of engineering network systems was mostly based on the planning design according to functional demand. To the network structural design of engineering systems, especially which is governed by disaster resistance performance, little attention has been paid. In the mid-1970s, the emergence of concept of lifeline earthquake engineering opened the first gap to break through this dilemma. In the following 40 years, starting from the seismic performance of buried pipelines and the earthquake disaster prediction of network systems, the researchers initiated a series of new research fields such as the vulnerability analysis of engineering structures, reliability analysis of engineering networks, optimization design of network topology and resilience analysis of engineering networks. These advances have formed the rudiment of a new branch of disciplines: engineering network reliability analysis and design.

The first author of this book has been involved in the research of lifeline engineering systems since the early 1990s. In the process of research, the author gradually formed such a basic concept: the disaster resistance design of engineering network system is an important part of engineering system design, and this kind of design should be based on the network reliability analysis. The author believes that the engineering network reliability analysis and design constitute an important hallmark of the development of civil engineering design theory in the new era. In fact, after a hundred years of developments, engineering reliability analysis and design has formed a complete theoretical framework: structural component reliability design—global structural reliability design—engineering network reliability design.

According to such a belief, this book is organized as the following four parts. After a brief introduction in Chap. 1, the first part of the book includes Chaps. 2 and

3, which introduce the seismic hazard analysis and advances in modeling of seismic ground motion, respectively. The second part, including Chaps. 4 and 5, describes the reliability analysis methods for structures in engineering network. Apart from the general reliability analysis methods, a global reliability analysis method of structure based on the probability density evolution theory is addressed briefly in this part. The seismic reliability analysis of a network system is the first key point of the book, which is the focus of the third parts and described in Chaps. 6 and 7. In this part, two reliability analysis methods of lifeline engineering networks, i.e. the connectivity reliability analysis and the functional reliability analysis, are introduced in detail, respectively. On the basis, the aseismic design and comprehensive control of composite lifeline system are addressed in the fourth parts of the book, including the topology optimization design methods for lifeline engineering networks, presented in Chap. 8, and the system control of composite lifeline engineering systems shown in Chap. 9, respectively. These contents constitute the fourth part of the book.

The book may be used as a textbook or research reference for graduate students and professionals in civil engineering. The level of the preparation assumed of the reader corresponds to that of the bachelor's degree in science or engineering.

The authors' sincere appreciations go firstly to Prof. Alfredo H-S Ang at the University of California, Irvine, and Prof. Pol. D. Spanos at Rice University, for their important advice in preparing the book and long-time friendship. Special thanks are also due to Prof. Bruce Ellingwood at Colorado State University, Prof. Dan Frangopol at Lehigh University, Prof. George Deodatis at Columbia University, Prof. Michael H. Faber at Aalborg University, Prof. Michael Beer at the Leibniz University of Hanover, Prof. Kok Kwang Phoon at the National University of Singapore and Prof. Yangang Zhao at Kanagawa University, for their valuable help and friendly encouragements. Taking this opportunity, the first author of the book would like to express his deep appreciation to his former students: Professor Jun He at Shanghai Jiaotong University, Prof. Lingli Chen at Shanghai University, Prof. Jianbing Chen and Prof. Wei Liu at Tongji University and Prof. Yuanfeng Bao at Xian Jiaotong University. Cooperation with them is always full of joy and inspiration. The authors are also indebted to their colleagues at Tongji University, Prof. Xilin Lu, Prof. Guoqiang Li, Prof. Yiyi Chen, Prof. Xianglin Gu and Prof. Qifeng Luo for their continuous cooperation and supports.

Finally, we would like to thank our families for their long-lasting support and love.

Shanghai, China
June 2020

Jie Li
Wei Liu

Contents

Chapter 1
Introduction

1.1 Lifeline Engineering Systems

Lifeline engineering systems, such as power systems, transportation systems, communication systems, water distribution systems, gas supply systems and so on, are essential to people's daily life. For modern society, such systems are the infrastructure systems supporting the function of modern city and regional economy.

As infrastructure systems, lifeline engineering systems are generally featured by owning three important characteristics. Firstly, they distribute in a large area, such as high-voltage electrical networks, regional transportation networks and urban water distribution networks, distribute in a large area. As a result, their performance is determined not only by the basic elements but also by the topology structures of the networks. Therefore, system analysis is important for the lifeline engineering systems to understand their performance comprehensively. Secondly, lifeline engineering systems are composed of different structural elements, i.e., engineering structures. For example, for the electrical power systems, there are generally buildings in power stations, high-voltage electrical equipment in transformer substations and high-voltage transmission towers in the transmission network. Similarly, roads and bridges in transportation systems and key buildings and equipment in communication systems are engineering structures as well. All these structures can be called as lifeline engineering structures. Their performance and operation states subjected to disasters determine whether the systems could maintain their functions or not. Thirdly, the performance of a lifeline engineering system directly or indirectly exerts influence on the other lifeline engineering systems. For example, whether electrical power systems work well has great impact on the operation of urban water distribution networks in the same area. Similarly, whether the transportation and oil transmission systems work well influences the operation of electrical power systems greatly. The interactions between different lifeline engineering systems become more apparent and extensive when serious disasters, such as strong earthquakes or hurricanes, happen. Considering the interactions between different lifeline engineering systems, a set of different lifeline systems can be regarded as a composite lifeline engineering system.

J. Li and W. Liu, *Lifeline Engineering Systems*,
https://doi.org/10.1007/978-981-15-9101-3_1

Certainly, other characteristics for different lifeline engineering systems can be found from different viewpoints. However, in this book only the structural characteristics of lifeline engineering systems are dealt with and therefore the other characteristics are not discussed.

The research on the disaster performance of lifeline engineering systems originated from the mid-1970s. In 1971, during the San Fernando earthquake, the infrastructure systems in San Francisco suffered serious damages. In the affected area, 11 power transmission lines and 47 highway bridges were destroyed, while more than 600 pipeline breaks appeared on water and gas systems. Meanwhile, the malfunctions of electrical power, water distribution and transportation systems after the earthquake attracted much attention (Duke & Moran, 1972). Based on the comprehensive investigation on this earthquake, the concept of lifeline earthquake engineering was presented and was accepted rapidly by the scientists and engineers (Duke & Moran, 1975). In 1974, a council, the Technical Council on Lifeline Earthquake Engineering (TCLEE), was founded in the American Society of Civil Engineers (ASCE). The first lifeline earthquake conference was organized by TCLEE in 1978 and the conference proceedings was published (Technical Council on lifeline Earthquake Engineering of ASCE, 1977). In 1983, a World Lifeline Earthquake Engineering Conference was origanized by the American Society of Mechanical Engineers (ASME) in Portland (Arima, 1983). In 1984, during the 8th World Conference on Earthquake Engineering in Japan, lifeline earthquake engineering was first listed as a session to exchange ideas between researchers and engineers. Since then, lifeline earthquake engineering has always been listed as a session in the later World Conference on Earthquake Engineering. Since 1990, Chinese, Japanese and American researchers on lifeline earthquake engineering met every 4 years to hold the series China-Japan-U.S. Trilateral Symposia on Lifeline Earthquake Engineering.

After nearly 50 years, the knowledge on lifeline engineering has been formulated based on a broader background. With the advancements of research and application, some new topics, including lifeline wind engineering, damage detection and health monitoring of lifeline systems, the resilience of lifeline system and so on, have appeared. Now, the research on lifeline engineering is becoming an important driving force for modern civil engineering.

Generally, lifeline engineering systems can be defined as the infrastructure systems which sustain proper operation and economic function of modern cities and communities (Li, 1999). This definition not only covers the previous research on urban lifeline engineering systems but also extends to regional lifeline engineering systems.

Considering the characteristic of research objective, a comprehensive investigation on lifeline engineering should include three levels, lifeline engineering structures, lifeline engineering networks and composite lifeline engineering systems. The research of lifeline engineering systems would provide scientific science methods and technical tools for modern lifeline engineering design considering performance in disaster scenavios and operational function in normal condition.

1.2 Damages of Lifeline Systems in Past Earthquakes

Among all kinds of natural disasters, strong earthquake is the biggest threat to lifeline engineering systems. During one strong earthquake, almost all kinds of lifeline engineering structures would suffer serious damages. Worse than that, part of or even all the functions of lifeline engineering systems will lose, including power cut, water supply stoppage and cell phone out of service. In some extreme cases, the whole system function will lose even if only some key structures suffer lightly or medium damages because of the fragility of lifeline engineering systems.

During many previous strong earthquakes, a lot of lifeline engineering systems suffered serious damages and lost their functions due to the strong earthquakes (Duke, 1971; Liu, 1986). Herein, the damages of some lifeline engineering systems are introduced under strong earthquakes in recent 30 years.

1. **American Loma Prieta Earthquake in 1989** ($M_L = 7.2$) (EERI, 1990)

During the American Loma Prieta Earthquake, taking place on October 17, 1989, the damage of the power system was a prominent phenomenon. The serious damages of 230 and 500 KV high-voltage substations resulted in that over 1.4 million customers encountered power cut (Fig. 1.1). In San Francisco, many robberies happened because of power cut. Meanwhile, over 350 pipeline breaks appeared in the water distribution network and more than 1000 gas leakages appeared in the gas supply network. Many oil tanks on soft soil ground were destroyed also.

Fig. 1.1 Damage of high-voltage substations

Fig. 1.2 Damage of bridges

2. **American Northridge Earthquake in 1994** ($M_L = 6.8$) (EERI, 1995)

During the American Northridge Earthquake, which happened on January 17, 1994, the transportation system was partially paralyzed due to the serious damages of some bridges (Fig. 1.2). Also, similar to the performance of the power system in American Loma Prieta Earthquake, many 230 and 500 KV high-voltage substations suffered serious damages. Moreover, a lot of high-voltage transmission towers collapsed due to soil liquefaction under the earthquake. As a result, more than 1.1 million customers encountered power cut. During the earthquake, 1400 pipeline breaks appeared in the water distribution network of Los Angeles. Meanwhile, over 150,000 gas leakages appeared in the gas supply network of Los Angeles, resulting in many fire accidents.

3. **Japanese Kobe Earthquake in 1995** ($M_L = 7.2$) (Investigation Group of Kobe Earthquake, 1997).

The epicenter of the Japanese Kobe Earthquake, which happened on January 17, 1995, located in a populous modern city. The investigation on the damages of lifeline engineering systems caused by this earthquake was the most comprehensive one in the history. During the earthquake, 5,438 people died and the direct economic loss was over 100 billion U.S. dollars. For the transportation system, six railroads crossing the earthquake areas suffered serious damages and many viaducts collapsed. Among 1,192 piers located in the highway from Osaka to Kobe, 611 piers were damaged and the damage rate was around 52%. Among those damaged piers, about 150 piers were destroyed and could not be repaired, which meant 13% piers should be rebuilt. Also, many damages appeared on the subway. For example, the tunnel roof of the Dakai subway station collapsed because the columns were damaged by shear force. After the earthquake, a 120 m long ground surface sank and the deepest location was over 3 m in the station area (Fig. 1.3). Also, Kobe Harbor was almost destroyed. About 80% embankment was destroyed and some cracks on embankment were 3 m deep. After the earthquake, one million customers encountered power cut and the repair process lasted 6 days. The damaged components of the power system included 48 275 and 77 KV substations and 446 power distribution circuits. The economic loss caused due to the damages of the substations and the distribution circuits was

Fig. 1.3 Surface subsidence due to the damages of subway station

over 55 billion Japanese Yen and 96 billion Japanese Yen, respectively. Among 1.36 million water customers in nine cities affected by the earthquake, including Kobe and Ashiya, 1.1 million customers encountered water stoppage due to 1,610 pipeline breaks in the water distribution network. A week later, only one third of the breaks were repaired. When the water distribution system recovered completely, two and a half months had passed. As a result, firefighting became very difficult due to lack of water. A newspaper report described the sad scene at that time, "firemen could do nothing but just see the fire burning with tears." Meanwhile, gas supply system suffered serious damages and 5,190 gas breaks appeared. Among these breaks, 109 breaks appeared on medium-pressure pipelines. After the earthquake, 857 thousand customers encountered no gas supply and the repair work lasted three months. For the communication systems, 3,170 communication lines were damaged. 19.7% communication lines located in the south of Hyogo and centered in Kobe were cut off due to the failures of telephone switchboards or the damages of communication lines. Many communication equipments were destroyed and some communication buildings were out of service (Fig. 1.4).

4. **Chinese Taiwan Chi-Chi Earthquake in 1999** ($M_L = 7.3$) (Architectural Institute in Taiwan, China, 1999)

During the Chinese Taiwan Chi-Chi Earthquake happened on September 21, 1999, 2,444 people died and 11,305 people were wounded. The economic loss reached 444 billion NT dollars. The transportation was disrupted in Central Taiwan, and many bridges, including Mingzu Bridge, collapsed. The railway stations of Central Taiwan and Chi-Chi suffered serious damages, including inflexion of rails and fall of train electrical lines. Meanwhile, due to the damages of the power system, 5.2 million customers covering a large area in the north and middle of Taiwan encountered power cut, resulting in a direct economic loss of 5.94 billion NT dollars. Also, the economic losses of water conservancy facilities and water pipelines were 4.72 billion and 950 million NT dollars, respectively.

Fig. 1.4 Damages of communication tower

5. **Chinese Wenchuan Earthquake in 2008** ($M_L = 8.0$) (State Key Laboratory of Disaster Reduction in Civil Engineering, Tongji University, 2008)

On May 12, 2008, a strong earthquake attacked Wenchuan in China. During the earthquake, the lifeline engineering systems suffered serious damages. In the affected area, 40 substations with 110 KV or higher voltage were damaged, resulting in that 180 transmission lines were out of service (Fig. 1.5). For gas supply systems, pipelines, equipments and other structures were damaged and the total number is as high as 51,000. For water distribution networks, 677 water factories and 138,000 km long pipelines were damaged in the affected area (Fig. 1.6). Taking the water distribution network of Mianzhu as an example breaks appeared on about 70% pipelines among 47 km-long mains pipelines, especially the gray cast iron pipelines and prestressed

Fig. 1.5 Damages of the power substation

Fig. 1.6 Crack of the pipe

cement pipelines. Because of the serious damages of the pipelines, a great deal of water leaked. When the water supply recovered on May 15, the leakage rate of the whole network was as high as 85%. Although the water supply increased from 35,000 tons per day before the earthquake to 50,000 tons per day after the earthquake, the water heads of many nodes were still very low, even zeros, resulting in the fact that water could not arrived at the third floor of many buildings.

6. **Great East Japan Earthquake in 2011** ($M_L = 8.0$) (Mimura et al., 2011)

On March 11, 2011, the Great East Japan earthquake happened. During the earthquake, the lifeline engineering systems suffered serious damages. In the affected area, reported damaged included 3,546 damages along roads. For gas supply systems, 20 breaks appeared on middle-pressure gas pipes, and 773 breaks appeared on low-pressure gas pipes. Meanwhile, low-pressure pipes inside 7,132 houses were damaged. For water distribution networks, a suspension of water supply occurred in about 2,300,000 houses in east Japan just after the earthquake. In Sendai City, the damage rate of pipes, defined as the number of damage divided by piping length, was 0.02 (/km).

1.3 Main Content of the Book

The main purpose of the book is to supply a basic theoretical framework and a technical tool for those who are interested in lifeline engineering systems.

There are totally nine chapters and three appendices in this book. In Chap. 2, the seismic hazard analysis is introduced. For regional lifeline engineering systems, such knowledge is important to evaluate and to calculate the seismic reliability of lifeline engineering networks. In Chap. 3, the research advancements on seismic ground motion are introduced. After introducing the general power spectrum model of ground motion briefly, the book focuses on the random ground motion field analysis based on physical background. In Chap. 4, the seismic analysis of buried pipeline is illustrated. Firstly, general seismic damage characteristics of buried pipelines are introduced briefly. Then, seismic response and seismic reliability analysis methods are introduced in detail. As buried pipelines are different in water distribution, gas supply and sewage systems, the suitability of these methods should be noted especially. In chapter 5, seismic analysis methods of engineering structures are presented. Besides the general seismic deterministic response analysis, a probability density evolution method for stochastic seismic response of structures is introduced and a global reliability analysis method for structures is described briefly. In Chaps. 6 and 7, two seismic reliability analysis methods of lifeline engineering networks, the connectivity reliability analysis and the functional reliability analysis, are introduced respectively. In Chap. 6, the seismic connectivity reliability analysis for general lifeline engineering networks is introduced. Particularly, a path-based and a cut-based recursive decomposition algorithm, which are suitable for large-scale networks, are presented. In Chap. 7, taking water distribution networks as an example, functional reliability analysis of lifeline engineering networks is introduced. In Chap. 8, the aseismic topology optimization design methods of lifeline engineering networks are presented. The basic idea of modern combinatorial optimization theory is introduced and its application to pipeline network is described in detail. In chapter 9, the seismic response simulation of composite lifeline systems and the system control based on simulation are introduced, respectively. These theories are related to the resilience of lifeline systems closely.

Three appendices are included in the book. In Appendix A, the basic knowledge of Boolean algebra is introduced. In Appendix B, the special analysis model for electrical power system is presented which is used in Chap. 6. In Appendix C, a useful seismic fire analysis model, which is used in Chap. 9, is introduced.

The authors believe that the contents described above have established the basic theoretical framework for seismic analysis and design of lifeline engineering systems.

References

Architectural Institute in Taiwan. (1999). *Investigation report on the great Jiji earthquake*. Architectural Institute in Taiwan (in Chinese).

Ariman, T. (1983). *Earthquake behavior and safety of oil and gas storage facilities, buried pipelines and equipment* (No. CONF-830607-). University of Tulsa.

Duke, C. M., & Moran, D. F. (1972). Earthquakes and city lifelines, San Fernando Earthquake of Feb.9, 1971 and public policy. *Joint committee on Seismic Safety of the California Legislature*, 53–67.

Duke, C. M., & Moran, D. F. (1975). Guidelines for evolution of lifeline earthquake engineering. In *U.S. National Conference on Earthquake Engineering* (pp. 367–376).

EERI. (1990). Loma Prieta earthquake reconnaissance report, *Earthquake Spectra*, Supplement to Vol.6.

EERI. (1995). Northridge earthquake of January 17, 1994 reconnaissance report. *Earthquake Spectra*, Supplement C to Vol.11.

Liu, H. X. (1986). *Damage in the great earthquake in Tangshan.* Seismological Press (in Chinese).

Li, J. (1999). Seismic response analysis and behavior control of complex lifeline engineering systems. *Bulletin of National Natural Science Foundation of China*, 13(6): 335–338 (in Chinese).

Investigation group of Japan Osaka great earthquake. (1997). Investigation Report on the Great Osaka Earthquake in Japan. *Seismological Press* (in Chinese).

Mimura, N., Yasuhara, K., Kawagoe, S., et al. (2011). Damage from the Great East Japan Earthquake and Tsunami - A quick report. *Mitigation Adaptation Strategies for Global Change, 16*(7), 803–818.

State Key Laboratory of Disaster Reduction in Civil Engineering. (2008). Earthquake Disaster of Wenchuan Earthquake, Shanghai: *Tongji University Press.*

Technical Conncil on lifeline Earthquake Engineering of ASCE. (1977). *The current state of knowledge of lifeline earthquake engineering.* New York: ASCE.

Chapter 2
Seismic Hazard Assessment

2.1 Introduction

Seismic hazard means the possibility of a strongest ground motion taking place in a specified area during a given time interval. Because earthquake occurrence in time, location and magnitude are random in nature, the probabilistic method is a reasonable choice for seismic hazard assessment.

For lifeline engineering systems, two basic concerns lie in seismic hazard assessment. Firstly, the seismic response analysis and reliability evaluation of lifeline engineering structures are based on seismic hazard assessment of the area where the structures are located. Secondly, as lifeline engineering systems usually cover a large area where the ground motion varies from site to site, therefore, the seismic hazard assessment would be helpful for the reliability evaluation of the whole system.

The research on seismic hazard assessment originated from the 1930s. Among the period between 1940 and 1960, the experimental method was used for seismic hazard assessment and the method is generally a deterministic method (Hu 1990; Zhang 1996). Because the earthquake occurrence in time, location and magnitude are all random in nature, the probabilistic method for seismic hazard assessment was introduced in the late 1960s. Cornell, an American scientist, was the pioneer who introduced random seismic hazard model systematically (Cornell 1968). As this model assumes that earthquake occurs from a point, it is called the point-source model. Because this model reflects the major aspects of the problem, several seismic zonation maps have been developed by using this model (ATC 1978; Hu 1990). In this chapter, the point-source model is introduced.

2.2 Uncertainty and Probability Model

The basic aim of seismic hazard assessment is to determine the probability distribution of the maximum intensity of ground motion at a specified site in a given time interval. To achieve this goal, a series of assumptions must be introduced to

J. Li and W. Liu, *Lifeline Engineering Systems*,
https://doi.org/10.1007/978-981-15-9101-3_2

reflect the randomness lying in earthquake occurrence time, location, magnitude and propagation path. In this way, the stationary Poisson process is adopted to delineate the randomness of earthquake occurrence time. The uncertainty of earthquake occurrence location is described by determining the potential seismic source and its classification. The magnitude-recurrence relationship is used to delineate the randomness of earthquake magnitude. The uncertainty of earthquake propagation path is expressed by the ground motion attenuation model, etc. The description of these models will be given as follows.

2.2.1 Earthquake Occurrence Probability Model

As mentioned above, the research on seismic hazard assessment began in the 1930s. Since then, many models have been presented (Savy et al. 1980; Nishioka and Shan 1980; Bender 1984). Despite the fruitful choices, there are two criteria to evaluate the practicability of the models. The first is whether the model could describe the earthquake occurrence sequence reasonably. The second is whether the basic statistical parameters of the model can be obtained from historic data. According to these two criteria, the Poisson process model is an ideal model.

A Poisson event is a random event having following three features.

1. Stationarity (homogeneity). In a special interval t, $t+\Delta t$, which can be time, area and so on, the event occurrence probability is proportional to Δt, while it does not dependent on the value of t. Herein, the event occurrence probability takes the value of $\lambda \cdot \Delta t$, where λ is the event occurrence rate. When the interval is time, the above feature is called stationarity. When the interval is cubage or area, it is called homogeneity.
2. Independence. Events can occur independently to each other in any interval and are not affected by the events in any other intervals.
3. Non repeatability. The probability that more than one event occurs in an interval is far less than $\lambda \cdot \Delta t$ and therefore can be neglected.

Existing research indicates that the occurrence time of earthquake sequence in one seismic zone is roughly consistent with the above features (Cornell 1968; Hu 1990). Moreover, when the earthquake sequence is classified into quiet and active periods, this consistency becomes more apparent. Therefore, Poisson process model can be used to delineate earthquake occurrence probability.

According to probability theory, the probability that an event with the above features occurs for n times in the time period T is.

$$P(n) = (\lambda T)^n e^{-\lambda T}/n! \quad (n = 0, 1, 2, \ldots) \tag{2.1}$$

Apparently, when the Poisson process model is adopted, the probability that an event occurs at an arbitrary time can be given with only one parameter λ.

Let v be the annual occurrence rate that an earthquake with magnitude $M \geq m$ occurs. According to Eq. (2.1), the probability that no earthquake with magnitude $M \geq m$ occurs in a time interval t is.

$$P(0) = e^{-\upsilon t} \quad (2.2)$$

Correspondingly, the probability that at least one earthquake with magnitude $M \geq m$ occurs in a time interval t is.

$$P(M \geq m) = 1 - P(0) = 1 - e^{-\nu t} \quad (2.3)$$

2.2.2 *Potential Seismic Zone*

Potential seismic zone is the zone where an earthquake may affect the specified site in a given time interval. Generally, the potential seismic zones are delineated for different areas, stripes or points with different maximum magnitudes to reflect the uncertainty of earthquake occurrence location. For one potential seismic zone, the probability of earthquake occurrence is assumed to be identical. Then, in one seismic zone, the number of earthquakes, $N(m)$, with magnitude larger than m can be obtained by historic earthquake records. Based on the assumptions mentioned above, the annual earthquake occurrence rate, ν_m, of a potential seismic zone can be stated as follows:

$$\nu_m = \frac{N(m)}{T} \quad (2.4)$$

where T is the earthquake recording time interval.

Obviously, T varies with different potential seismic zones. Also, for one potential seismic zone, $\nu_m$varies with m. Apparently, for a specified ν_m, the probability that at least one earthquake with magnitude $M \geq m$ occurs in the potential seismic zone in a given time interval, t, can be given by Eq. (2.3).

Considering the seismic risk of a specified site, the division of potential seismic source can be determined by synthesizing the information in terms of the seismic geotechnical structures, historic earthquake records, modern seismic activity records and geophysical field distribution in a specified investigation area. Figure 2.1 gives a general program to divide potential seismic zone.

The scope of investigation area should cover an area centered in the target engineering site with 250–300 km in radius. Figure 2.2 indicates a distribution of potential seismic zone in North China.

For a potential seismic source with few historic earthquake records, it seems impossible to give the value of ν_m. An alternative method is to calculate it according to the historic earthquake records of the whole seismic zone which covers the specified

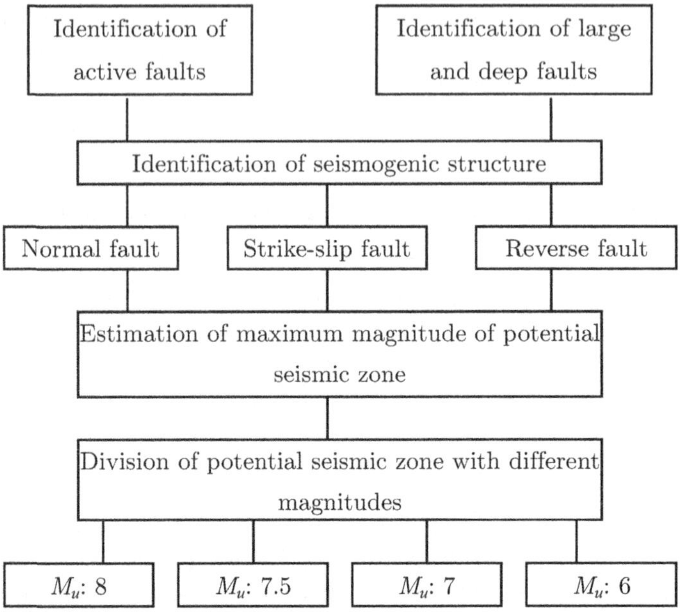

Fig. 2.1 Program of potential seismic zone division

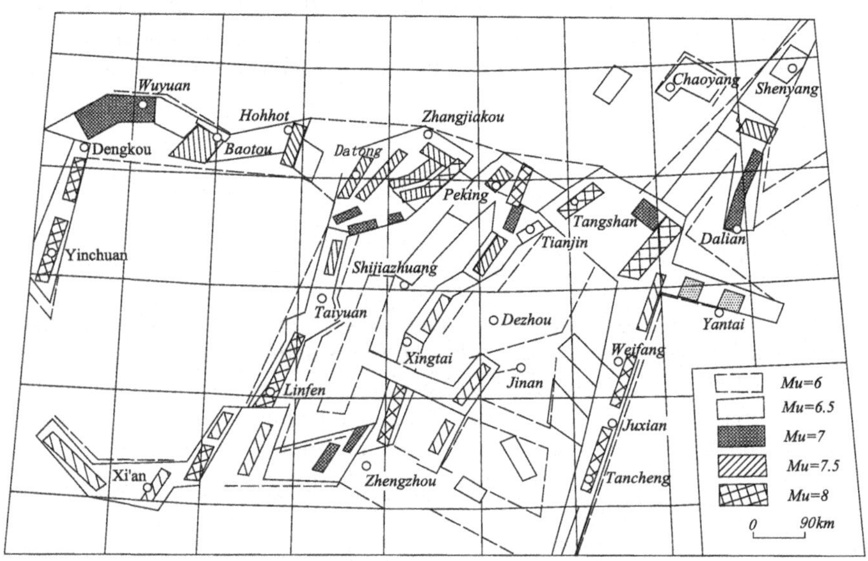

Fig. 2.2 Distribution of potential seismic source zone with magnitudes larger than 6 in North China

site. Then the annual occurrence rate of *j*th potential seismic zone is (Hu 1990)

$$\nu_{mj} = \frac{S_j}{S}\nu_m \tag{2.5}$$

where S_j is the area of *j*th potential seismic zone, *S* is the area of the whole seismic zone and ν_m is the annual earthquake occurrence rate of the whole seismic zone.

2.2.3 Probability Distribution Function of Earthquake Magnitude

Usually, the occurrence probability of earthquakes with high magnitude is relatively low, while that of earthquakes with low magnitude is relatively high. A magnitude-frequency relationship, which indicates the probability distribution characteristic of earthquake magnitude in an area, can be used to estimate the occurrence frequency of earthquakes with different magnitudes.

Let $N(M)$ be the frequency of earthquake with magnitudes greater than M for a specified zone. The historic earthquake records in different zones indicate that the natural logarithm of N is approximately a linear function of M, that is.

$$\ln N(M) = a - bM \quad M_0 \le M \le M_u \tag{2.6}$$

where a and b are empirical coefficients and vary with zones, M_0 is the lower boundary of magnitude and generally takes the value of 4–4.75, M_u is the upper boundary of magnitude and determined by the earthquake activity of this zone (Hu 1990).

Usually, Eq. (2.6) is called the Gutenberg-Richard equation (Gutenberg and Richter 1941). A typical magnitude-frequency relationship is shown in Fig. 2.3.

Obviously, $N(M)$ can be derived from Eq. (2.6) as follows:

$$N(M) = e^{a-bM} \quad M_0 \le M \le M_u \tag{2.7}$$

Therefore, the cumulative distribution function and the probability density function of earthquake magnitude can be given easily. In fact, the cumulative distribution function of earthquake magnitude means the occurrence probability of earthquakes with the magnitude smaller than the value *m*, that is

$$\begin{aligned} F(m) &= P(M \le m) \\ &= \frac{N(M_0) - N(m)}{N(M_0) - N(M_u)} \end{aligned}$$

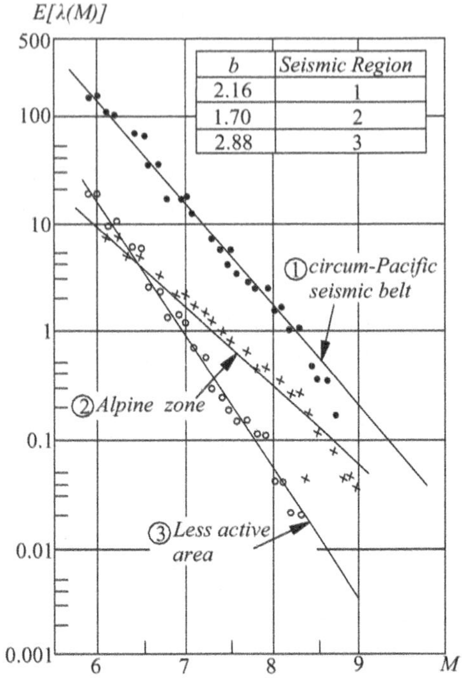

Fig. 2.3 A typical magnitude-frequency relationship

$$= \frac{1 - e^{-b(m - M_0)}}{1 - e^{-b(M_u - M_0)}} \tag{2.8}$$

Correspondingly, the probability density function is the derivation of $F(m)$.

$$f(m) = \frac{be^{-b(m - M_0)}}{1 - e^{-b(M_u - M_0)}} \tag{2.9}$$

Apparently, the magnitude-frequency relationship and probability distribution function of earthquake magnitude is equivalent and they all delineate the randomness of earthquake magnitude.

2.2.4 *Ground Motion Attenuation*

The energy carried by seismic wave decreases gradually with the increase of distance from the seismic source. Consequently, ground motion intensity attenuates with the increase of propagation distance. This attenuation relationship exhibits apparent variability for different media and sites. Therefore, a simplification method should be adopted to clarify the uncertainty of earthquake propagation path. In seismic hazard analysis, a mean attenuation relationship of ground motion on bedrock or site with middle stiffness condition is used to delineate the influence of propagation path.

A general ground motion attenuation relationship can be stated as follows (Trifunac 1976; Hu 1990):

$$lnY = a + bM + cln(R_h + R_0) + dT_s \tag{2.10}$$

where *Y* is the ground motion parameter, such as the amplitude of seismic acceleration or seismic velocity, *a*, *b*, *c* and *d* are empirical coefficients, *M* is the earthquake magnitude, R_h is the distance between seismic source and specified site, R_0 is a constant related to the depth of seismic source and T_s is the natural period of site.

According to (2.10), the ground motion attenuation relationship can be rewritten as follows:

$$Y = b_1(R_h + R_0)^{b_3} e^{(b_2 M + b_4 T_s)} \tag{2.11}$$

Apparently, the relationships of the empirical coefficients between Eqs. (2.10) and (2.11) are:

$$b_1 = e^a, b_2 = b, b_3 = c, b_4 = d \tag{2.12}$$

For the seismic wave propagation in bedrock, T_s has little effect and can be neglected. That is:

$$Y = b_1(R_h + R_0)^{b_3} e^{b_2 M} \tag{2.13}$$

Some typical ground motion attenuation relationships are shown in Table 2.1.

2.3 Seismic Hazard Analysis Method

Seismic hazard analysis can be defined as to calculate the exceedance probability of ground motion at a specific site. Generally, the peak acceleration ($y = a_{max}$) is taken as an identified variable of ground motion. According to the knowledge about seismic sources, there are three different analytical models: point-source model, line-source model and area-source model.

2.3.1 Point-Source Model

If the location of seismic source is known and can be considered as a point source (Fig. 2.4), then the mean annual rate that earthquake occurs in a unit area is:

Table 2.1 Some typical ground motion attenuation relationships (Li, 2005)

	Attenuation Relationship
Amplitude	$a_{\max} = 0.836e^{0.86M}(R+25)^{-1.35}$ $v_{\max} = 8.248e^{1.07M}(R+25)^{-1.55}$
	$a_{max} = \begin{cases} 46 \times 10^{0.208M}(R+10)^{-0.686} & Bedrock \\ 24.5 \times 10^{0.333M}(R+10)^{-0.924} & Solid\ Base \\ 59 \times 10^{0.261M}(R+10)^{-0.886} & Inferior - solid\ Base \\ 12.8 \times 10^{0.482M}(R+10)^{-1.112} & Soft\ Base \end{cases}$
	$a_{\max} = 3.79 \times 10^{-3}e^{1.15M}D^{-0.83}e^{-0.00159}$ $v_{\max} = 2.51 \times 10^{-4}e^{2.3M}D^{-0.83}e^{-0.00076R}$ $D = (R^2 + H_{\min})^{0.5} H_{\min} = 0.0186e^{1.05M}$
	$a_{\max} = 214.5e^{0.819M}(R+15)^{-1.639}$(North China) $a_{\max} = 773e^{0.931M}(R+20)^{-2.005}$ (West America)
Duration	$\ln T_d = 0.19 + 0.15M + 0.35 \ln R + 0.735 + 0.23V$ $S = \begin{cases} 0 \text{ Bedrock} \\ 1 \text{ Soil} \end{cases}$ $V = 0$ Horizontal $r = 1$ Vertical

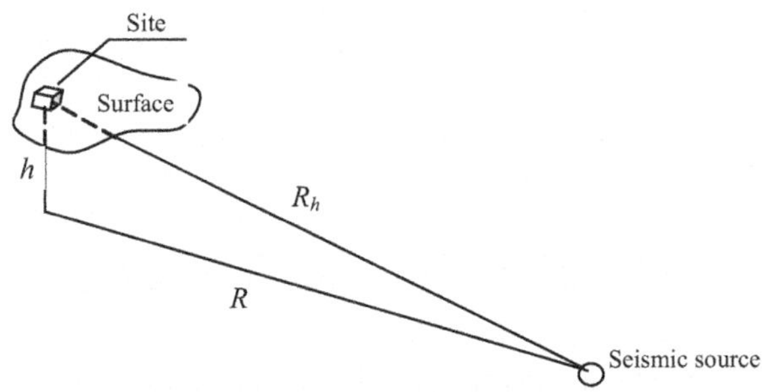

Fig. 2.4 Illustration of the point source model

$$\gamma = \frac{N(m)}{A \cdot T} \tag{2.14}$$

where A is the area of potential seismic source zone, $N(m)$ is the number of earthquakes that exceeds a given magnitude m, T is the period of earthquake records.

$N(m)$ can be obtained by the magnitude-frequency relationship in Eq. (2.7). Assume that:

$$\alpha = a - \ln(AT) \tag{2.15a}$$

$$\beta = b \tag{2.15b}$$

where a and b are the empirical coefficients in Eq. (2.7).

Then, according to Eqs. (2.7) and (2.14), the mean annual rate can be derived:

$$\gamma = e^{\alpha - \beta m} \tag{2.16}$$

Considering the acceleration attenuation relationship in Eq. (2.13), for a given distance R_h, the corresponding earthquake magnitude is

$$M = \ln\left[\frac{Y}{b_1}(R_h + R_0)^{-b_3}\right]^{\frac{1}{b_2}} \tag{2.17}$$

Therefore, for a given peak acceleration y, the corresponding earthquake magnitude is

$$m = \ln\left[\frac{Y}{b_1}(R_h + R_0)^{-b_3}\right]^{\frac{1}{b_2}} \tag{2.18}$$

Consequently, the peak acceleration Y at a specified site is greater than a given peak acceleration y means that the corresponding earthquake magnitude $M \geq m$. That is.

$$P(Y \geq y) = P(M \geq m) \tag{2.19}$$

According to the earthquake occurrence model in Eqs. (2.3) and (2.14), the probability that earthquake with magnitude $M \geq m$ occurs at least once in t years is

$$P_t(M \geq m) = 1 - e^{-\gamma t} \tag{2.20}$$

From Eqs. (2.10) and (2.20), there exists

$$P(Y \geq y) = 1 - e^{-\gamma t} \tag{2.21}$$

Noticing Eqs. (2.16) and (2.18), the mean annual rate in the above equation can be described as follows:

$$\gamma = e^{\alpha}\left(\frac{y}{b_1}\right)^{-\frac{\beta}{b_2}}(R_h + R_0)^{\frac{\beta b_3}{b_2}} \tag{2.22}$$

2.3.2 Line-Source Model

When the location and direction of seismic fault are known, the potential seismic source can be viewed as a line-source (Fig. 2.5).

Herein, the line-source with L in length can be divided into k segments with the length Δl_i ($i = 1, 2, \ldots, k$). Each segment can be regarded as a point-source and the occurrence of earthquake in each segment is assumed mutually independent to each other.

Let $P(E_{l_i})$ represents the probability that no earthquake with magnitude $M \geq m$ occurs on the ith segment in t years. As the event that no earthquake occurs on the line-source is equivalent to the event that no earthquake occurs in any segment, the probability that no earthquake occurs on the line-source in t years is

$$P_t(0) = \prod_{i=1}^{k} P(E_{l_i}) = \prod_{i=1}^{k} \mathrm{e}^{-\gamma.\Delta l_i t} = \mathrm{e}^{-\sum_{i=1}^{k} -\gamma \cdot \Delta l_i t} \tag{2.23}$$

With the increment of k, the length of each segment approaches to zero. Then, the probability that no earthquake with magnitude $M \geq m$ occurs on the line-source becomes

$$P_t(0) = \lim_{\Delta l \to 0} \exp\left[-\sum_{i=1}^{k} \gamma \cdot \Delta l_i \cdot t\right] = \exp\left[-\int \gamma dl \cdot t\right] \tag{2.24}$$

Therefore, the probability that at least one earthquake with magnitude $M \geq m$ occurs on the line-source in t years is:

$$P(M \geq m) = 1 - P_0(0) = 1 - \exp\left[-\int \gamma \mathrm{d}l \cdot t\right] \tag{2.25}$$

According to Eq. (2.19), the probability that the ground motion $Y \geq y$ occurs at least once in the specified site in t years is:

$$P_t(Y \geq y) = 1 - e^{-\gamma_L t} \tag{2.26}$$

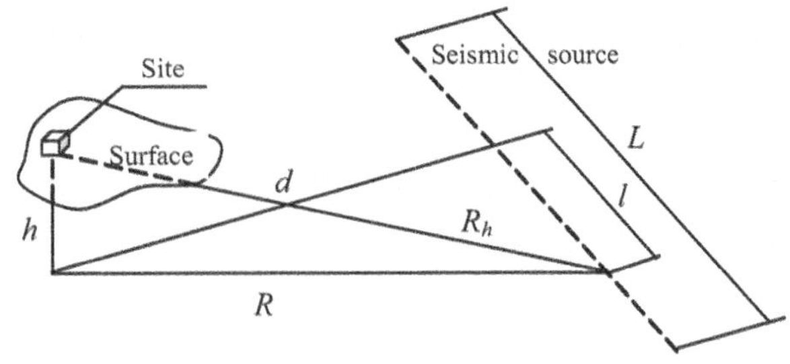

Fig. 2.5 Illustration of a line-source model

where.

$$\gamma_L = e^{\alpha}\left(\frac{y}{b_1}\right)^{-\frac{\beta}{b_2}} \int (R_h + R_0)^{\frac{\beta b_3}{b_2}} dl \tag{2.27}$$

Apparently, the only difference between the exceedance probability of the point-source model and that of the line-source model is their mean annual rate. The mean annual rate of line-source is equal to the integral of that of its points along the line.

2.3.3 Area-Source Model

When only the distribution area of seismic fault is known, the potential seismic source should be regarded as an area-source. Similar to the line-source model, the area-source can be divided into a series of small area elements. The probability that at least one earthquake with ground motion $Y \geq y$ occurs in the specified site is.

$$P_t(Y \geq y) = 1 - e^{-\gamma_A t} \tag{2.28}$$

where.

$$\gamma_A = e^{\alpha}\left(\frac{y}{b_1}\right)^{-\frac{\beta}{b_2}} \cdot \theta \cdot \int (R_h + R_0)^{\frac{\beta b_3}{b_2}} R dR \tag{2.29}$$

where R_1 and R_2 are shown in Fig. 2.6. Herein, the polar coordinate system is introduced to simplify the analysis.

2.3.4 Probability Distribution Function of Ground Motion Amplitude

If there are N_{p} point-sources, N_{l} line-sources and N_{a} area-sources in a specified region, then the probability that at least one earthquake with ground motion $Y \geq y$ occurs in T years is

$$P(Y \geq y) = 1 - \exp\left\{-\left(\sum_{i=1}^{N_p} \gamma_i + \sum_{j=1}^{N_l} \gamma_{lj} + \sum_{k=1}^{N_a} \gamma_{Ak}\right) T\right\} \tag{2.30}$$

Equation (2.30) indicates that the exceedance probability of ground motion is the accumulation of the contribution of different potential seismic sources. In a specified site, the more potential seismic sources, the higher the exceedance probability of ground motion.

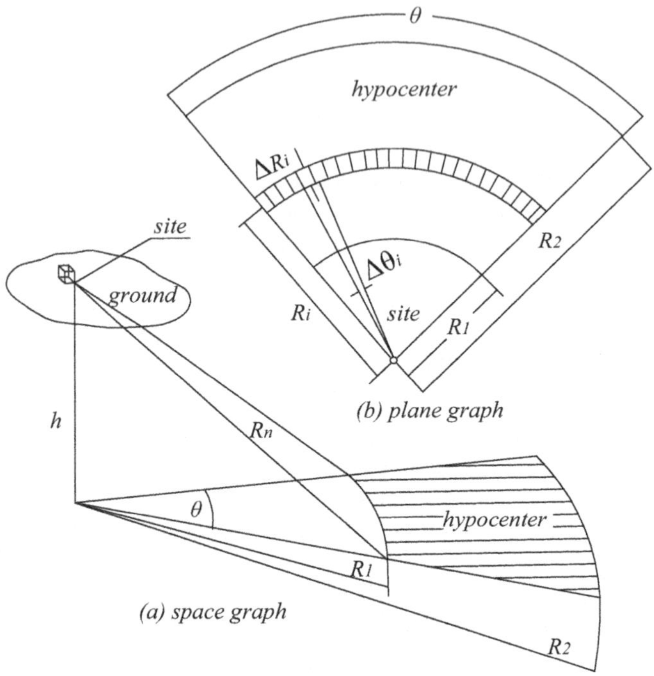

Fig. 2.6 Illustration of an area-source model

The cumulative probability function of the ground motion amplitude (or maximum of ground motions) can be stated as follows:

$$
\begin{aligned}
F_t(y) &= P_t(Y < y) = 1 - P_t(Y \geq y) \\
&= \exp\left\{-\left(\sum_{i=1}^{N_p} \gamma_i + \sum_{j=1}^{N_l} \gamma_{lj} + \sum_{k=1}^{N_a} \gamma_{Ak}\right) T\right\}
\end{aligned} \tag{2.31}
$$

Let $\gamma_\alpha = \gamma \cdot y^{\frac{\beta}{b_2}}$, then the cumulative probability function becomes

$$
F_t(y) = \exp\left\{-y^{-\frac{\beta}{b_2}} \cdot v_\alpha \cdot T\right\} = \exp\left(-\frac{\tau}{y}\right)^{\lambda} \tag{2.32}
$$

where.

$$
v_\alpha = \sum_{i=1}^{N_p} \gamma_i + \sum_{j=1}^{N_l} \gamma_{lj} + \sum_{k=1}^{N_a} \gamma_{ak} \tag{2.33}
$$

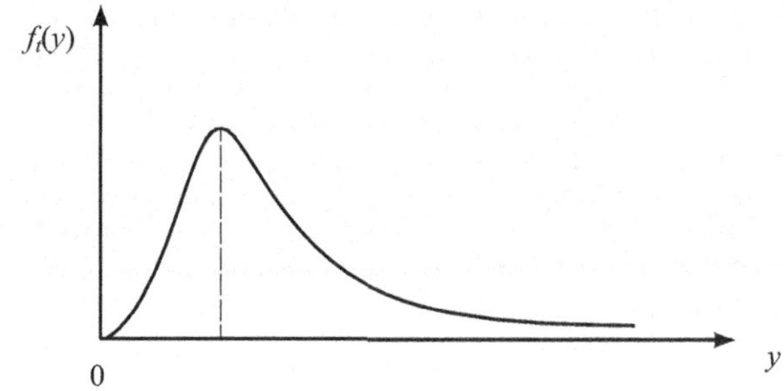

Fig. 2.7 A typical extreme II distribution

$$\lambda = \frac{\beta}{b_2} \tag{2.34}$$

$$\tau = (v_\alpha T)^{\frac{b_2}{\beta}} \tag{2.35}$$

Equation (2.32) indicates an extreme II distribution and its probability density function is.

$$f_t(y) = -y^{-(1+\lambda)}\tau^{\lambda}\exp\left(-\frac{\tau}{y}\right)^{\lambda} \tag{2.36}$$

Figure 2.7 indicates a typical result of such distribution. The function actually gives the probability density function of the seismic acceleration amplitude.

It is worth pointing out that, in Eqs. (2.32)–(2.36), the coefficient β for different potential seismic sources is assumed to be the same value, so is for the ground motion attenuation coefficient, b_2. These assumptions are used only to indicate that the probability distribution of seismic peak acceleration is an extreme II distribution. In general condition, the probability density function of ground motion should be derived from Eq. (2.31) by the differentiation of $F(y)$ in term of y.

References

ATC-40, (1978), Seismic Retrofit Practices Improvement Program: Seismic Evaluation and Retrofit of Concrete Buildings. *Seismic Safety Commission*, State of California.

Bender, B. (1984). A two-state Poisson model for seismic hazard estimation. *Bulletin of the Seismological Society of America, 74*(4), 1463–1468.

Cornell, C. A. (1968). Engineering seismic risk analysis. *Bulletin of the Seismological Society of America, 58*(5), 1583–1606.

Gutenberg, B., & Richter, C. F. (1941). *Seismicity of the earth* (p. 34). Special Papers No: Geological Society of America.

Hu, Y.X. (1990). Comprehensive Probability Method for Seismic Hazard Analysis. *Seismological Press* (in Chinese).

Li, J. (2005). Lifeline earthquake engineering-basic method and application. Beijing: Science Press (in Chinese).

Nishioka, T. C., & Shan, H., (1980). Application of the Markor chain on probability of earthquake occurrences. In *Japan Society of Civil Engineers* (pp. 137–145).

Savy, J. B., Boore, D., & Shah, H. C. (1980). Nonstationary risk model with geaphysical input. *Journal of the Structural Division, 106*(1), 145–163.

Trifunac. (1976). Preliminary empirical model for scaling fourier amplitude spectra of strong ground acceleration in terms of earthquake magnitude, source-to-station distance, and recording site conditions. *Bulletin of the Seismological Society of America*, 66(4): 1343–1373.

Zhang, Z. Y. (1996). *Seismic hazard analysis and its application.* Tongji University Press (in Chinese).

Chapter 3
Seismic Ground Motion Model

3.1 Introduction

When an earthquake happens, the seismic wave produced by the seismic source is a time-dependent process. Through the propagation in the earth media, the wave shape will undergo complex changes. For a given site, the ground motion can be characterized by the time history of ground motion displacements, velocities or accelerations. Due to the influence of a series of uncontrollable factors like the mechanism of the seismic source, the earthquake propagation paths and the geotechnical media distribution in the engineering site, the ground motion is a typical stochastic process and varies in the spatial location. Therefore, the spatial random field model should be adopted as the basic model to represent the seismic ground motion. However, due to the difficulties in modeling, some simplifications may be introduced in the modeling process.

One of the most common simplifications is to neglect the spatial variation of the ground motion, and correspondingly the spatial random fields would be a series of random processes with the same statistical characteristics. That is, to use a random time sequence at one point, and to assume the ground motion at various points being identical within the acceptable range. For example, considering the ground motion input into the bottom of building structures with a relatively small planar size and assuming the building base within the scope of the foundation as a rigid plate, the input of ground motion can be reflected by a stochastic process model at one point.

If the difference among the ground motions acting on different points of the bottom of structure cannot be neglected, the random field model should be adopted to reflect the spatial variation of ground motions. In these cases, the uniform and isotropic assumptions may be used to simplify the model. In the uniform assumption, the difference among the points in the random fields is only related to the distance between two points and had nothing to do with the location, while according to the isotropic assumption, the probability distribution of the random fields is independent of the orientation.

J. Li and W. Liu, *Lifeline Engineering Systems*,
https://doi.org/10.1007/978-981-15-9101-3_3

There are two basic methods for modeling seismic ground motion: phenomenology-based modeling and physically-based modeling. Due to the difficulties in modeling of the multidimensional probability distribution function, the correlation function or the power spectral density function is used commonly in the phenomenology-based modeling. In essential, the method is based on statistical moments and therefore is called statistically-based modeling. In contrast, the physically-based modeling try to give a random function model of ground motions with the real physical background. The later can not only give a complete mathematical description for the stochastic processes or random fields but also could provide a feasible manner for the experimental verification of the stochastic processes or random fields possible because of their physical significance.

This chapter will outline the phenomenological models and physical models for the seismic ground motion. Usually, for the strong ground motions, the reliable observation data are ground acceleration. Therefore, the seismic ground motion model often refers to the acceleration model.

3.2 Statistically-Based Model

3.2.1 Stationary and Non-stationary Processes

For a stationary process, particularly if it can be regarded as a Gaussian process, the statistical characteristics of the stochastic process can be totally determined when the mean and the correlation function or the power spectral density function are known. The mean of a stationary processes $X(t)$ is a constant, and the correlation function is a function of the time interval $\tau = t_2 - t_1$, that is

$$m_X(t) = c \tag{3.1}$$

$$R_X(\tau) = R_X(t_2 - t_1) \tag{3.2}$$

If $c = 0$, the correlation function $R_x(\tau)$ and the power spectral density function $S_x(\omega)$ have the following relationship:

$$S_X(\omega) = \int_{-\infty}^{+\infty} R_X(\tau)e^{-i\omega\tau}d\tau \tag{3.3}$$

$$R_X(\tau) = \frac{1}{2\pi}\int_{-\infty}^{+\infty} S_X(\omega)e^{-i\omega\tau}d\omega \tag{3.4}$$

It is indicated that $S_X(\omega)$ and $R_X(\tau)$ consist of a Fourier transform pair.

To determine a specific stationary process model, the ergodicity assumption is generally introduced. It is indicated that, for an ergodic process, various random behaviors could appear in a long enough time series. Therefore, the assemble average can be replaced by the time average. For an ergodic process, the estimated values of the mean and the auto-correlation function can be obtained using the measured samples as follows

$$\widehat{m}_X = \frac{1}{N}\sum_{i=1}^{N} X_i \tag{3.5}$$

$$\widehat{R}_X(\tau_k = k) = \frac{1}{N-k}\sum_{i=1}^{N-k} X_i X_{i+k} \tag{3.6}$$

where $X_i = X(t_i)$ is the amplitude of the sampled process at the time instant t_i, k is the lag number, N is the total number of the sampled points, the hat $\hat{\cdot}$ represents the estimated value of the variable.

The power spectral density function can be obtained from the discrete Fourier transform of $R_X(\tau_k)$ by using Eq. (3.3). However, such spectral estimation results are usually biased. To obtain the unbiased results, the maximum entropy spectral method can be adopted (Burg 1967).

If a series of measured samples are available, the sample set should be used to achieve the estimation of the power spectral density. In this case, the auto-power spectral density is given as follows:

$$\widehat{S}_X(\omega) = \frac{1}{M}\sum_{i=1}^{M}\left[\frac{1}{T}|X_i(\omega)|^2\right] \tag{3.7}$$

where M is the sample number of the sample set, T is the observation time duration, $X_i(\omega)$ is the Fourier spectrum of the sample process.

A stationary process is a scientifically abstract model. Most of the actual ground motion does not possess the stationary characteristics. Therefore, it is necessary to adopt the non-stationary process to establish the model of seismic ground motion.

For non-stationary stochastic processes, the ergodicity assumption will be no longer valid. Therefore, the assemble averages cannot be replaced by the time averages. In other words, the model should be established based on the sample sets. In practice, a type of uniformly modulated non-stationary process models is usually employed to build a non-stationary random excitation model (Clough and Penzien, 1993). This model can be expressed as a product of a deterministic time function $f(t)$ and a stationary process $X_s(t)$, namely

$$X(t) = f(t)X_s(t) \tag{3.8}$$

Here, the mean value of $X(t)$ is zero.

When the correlation function or the power spectral density function of $X_s(t)$ is given, it is easy to calculate the correlation function or the power spectral density function of the non-stationary random excitation $X(t)$, that is

$$R_X(t_1, t_2) = f(t_1)f(t_2)R_{X_S}(\tau) \tag{3.9}$$

$$S_X(t, \omega) = f^2(t)S_{X_S}(\omega) \tag{3.10}$$

To get the model of the non-stationary random excitations, it is required first to isolate $f(t)$ from the sample functions. For a time process sample $x_i(t)$, it can be done by a variety of approaches. For example, we can first define

$$y_i(t) = |x_i(t)| \tag{3.11}$$

and then use the technique employed in the Empirical Mode Decomposition to determine the upper and lower envelope curves in the sifting process (Huang et al. 1998) to specify the upper envelope $f_i(t)$ of $y_i(t)$.

For a sample set, the specific expression of $f(t)$ can be determined by assuming the function form of $f(t)$ and using the least squares method to fit the function through $f(t)$ $(i = 1, 2, \ldots, \text{n})$.

Once $f(t)$ is determined, and the sample set of the stationary process $X_s(t)$ is derived, then the modeling methods on stationary random processes can be used to complete the modeling of $X_s(t)$ and the non-stationary excitations models can be obtained by Eq. (3.8).

3.2.2 *One-Dimensional Stochastic Process Model*

The actual earthquake records show that the time process of the ground motion accelerations usually includes three stages of vibrations: the initial, the strong and the attenuating stages (Fig. 3.1). Therefore, the ground motion is a typical non-stationary process. When the stationary process model is used to establish the ground motion models, it is usually believed that this only reflects its strong motion stage.

For simplification, the ground motion on the engineering site surface may be regarded as a filtered white noise. Following this consideration, if the ground motion on the bedrock is assumed as a zero-mean white noise process with the spectral density S_0, and the soil surface is simulated as a single-degree-of-freedom linear system, the Kanai-Tajimi spectrum model can be obtained (Kanai 1957; Tajimi 1960):

$$S(\omega) = \frac{1 + 4\zeta^2\left(\frac{\omega}{\omega_0}\right)^2}{\left[1 - \left(\frac{\omega}{\omega_0}\right)^2\right]^2 + 4\zeta^2\left(\frac{\omega}{\omega_0}\right)^2} S_0 \tag{3.12}$$

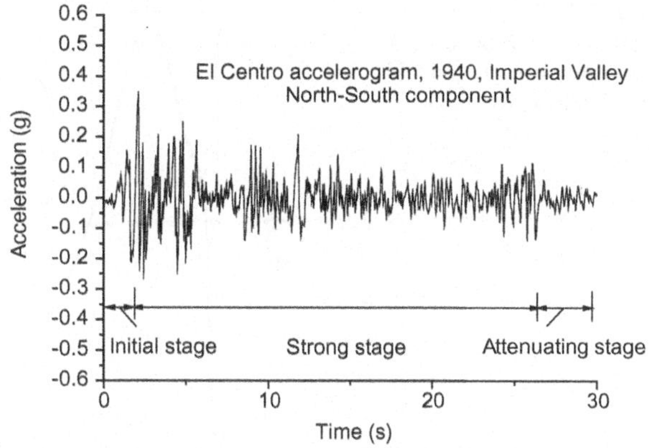

Fig. 3.1 Three stages of a typical ground motion record

where $S(\omega)$ is the power spectral density function of the stationary ground motion process, ζ is the damping ratio of the soil on the site and ω_0 is the natural frequency of the site.

The physical meaning of the model above is clear that the influence of the soil property on the frequency spectrum of ground motion is considered. However, this model inappropriately exaggerates the low-frequency content of the ground motion. Meanwhile, the ground motion velocity and displacement obtained from the model have a singular point where the frequency is zero, therefore, a finite variance of the ground motion displacement and velocity cannot be calculated. To overcome these shortcomings, the following correction model has been introduced (Hu and Zhou 1962):

$$S(\omega) = \frac{1 + 4\zeta^2\left(\frac{\omega}{\omega_0}\right)^2}{\left[1 - \left(\frac{\omega}{\omega_0}\right)^2\right]^2 + 4\zeta^2\left(\frac{\omega}{\omega_0}\right)^2} \cdot \frac{\omega_c^n}{\omega_c^n + \omega_0^2} S_0 \tag{3.13}$$

where ω_c is the low-frequency decrease factor, and $n = 4$–6.

There are many similar correction models, for example, adding a filter to the model (3.12) forms a double white noise process (Ruiz and Penzien 1969):

$$S(\omega) = \frac{1 + 4\zeta^2\left(\frac{\omega}{\omega_0}\right)^2}{\left[1 - \left(\frac{\omega}{\omega_0}\right)^2\right]^2 + 4\zeta^2\left(\frac{\omega}{\omega_0}\right)^2} \cdot \frac{\left(\frac{\omega}{\omega_1}\right)^4}{\left[1 - \left(\frac{\omega}{\omega_1}\right)^2\right]^2 + 4\zeta_1^2\left(\frac{\omega}{\omega_1}\right)^2} S_0 \tag{3.14}$$

where ζ_1 and ω_1 are the parameters of the assumed second filter.

In contrast to Model (3.12) in which the power spectral density is a finite value different from zero at $\omega = 0$, both Models (3.13) and (3.14) can ensure the power spectral density to be zero at $\omega = 0$ (Fig. 3.2).

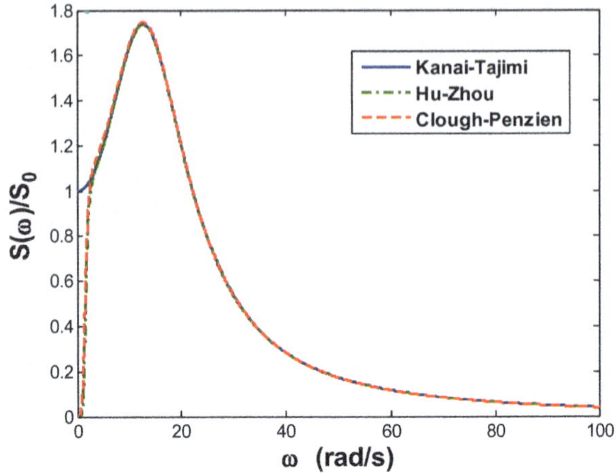

Fig. 3.2 Comparison of different PSD models (**Parameters**: $\omega_0 = 15.71$, $\omega_c = 0.1\omega_0, \zeta = 0.64$, $\omega_1 = 0.1\omega_0, \zeta_1 = \zeta$)

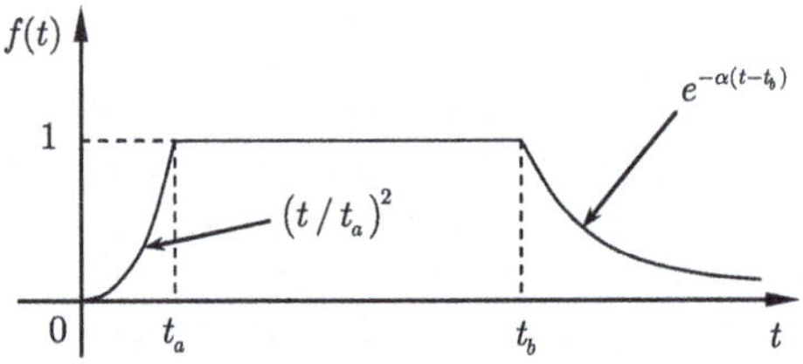

Fig. 3.3 Envelop function for ground motion

If it is needed to reflect the rising and decaying sections of the ground motion process, that is, to reflect the non-stationary nature of the ground motion, a modulate non-stationary stochastic process model can be introduced [cf. Equation (3.8)]. The modulate envelope function can be given by, say (Amin and Ang 1968; Jennings et al. 1968)

$$
\mathrm{f}(t) = \begin{cases} (t/t_a)^2, & \text{for } t \le t_a \\ 1, & \text{for } t_a < t \le t_b \\ e^{-\alpha(t-t_b)}, & \text{for } t \ge t_b \end{cases} \tag{3.15}
$$

where t_a and t_b are, respectively, the starting time and the end time of the stationary section of the strong ground motion, α is a parameter controlling the decaying speed of the attenuation section (Fig. 3.3).

3.2.3 *Random Field Model*

When the difference between the ground motions at two points with a certain distance cannot be neglected, it is required to use the random fields to describe the ground motion. Using the spatial discretization method (VanMarke, 1983), the continuous random field description can be transformed into a set of stochastic processes. Therefore, a uniform and isotropic random field B can be represented by the following power spectral density matrix:

$$S_B(\omega) = \begin{bmatrix} S_{11}(\omega), & S_{12}(\omega), & \cdots, & S_{1m}(\omega), \\ S_{21}(\omega), & S_{22}(\omega), & \cdots, & S_{2m}(\omega), \\ \ldots, & \ldots, & \ldots, & \ldots, \\ S_{m1}(\omega), & S_{m2}(\omega), & \cdots, & S_{mm}(\omega), \end{bmatrix} \tag{3.16}$$

where m is the number of spatial points; $S_{kj}(\omega)$ is the cross-spectral density, which is a complex function and characterizes the correlation degree between the stochastic ground motions at the points k and j. If $k = j$, then it is the auto-spectral density of the ground motion at one point.

In the study of ground motion, a coherence function is usually used which usually represents the correlation feature between the ground motion at two different points, and is defined as follows:

$$\gamma_{kj}(\omega) = \begin{cases} \frac{S_{kj}(\omega)}{\sqrt{S_{kk}(\omega)S_{jj}(\omega)}}, & if\ S_{kk}(\omega)S_{jj}(\omega) \neq 0 \\ 0, & otherwise \end{cases} \tag{3.17}$$

Generally, for a small-scale engineering site, the power spectrum density of each point can be assumed to own the same value, i.e., $S(\omega)$. Then, Eq. (3.16) can be rewritten as follows:

$$S_B(\omega) = S(\omega) \begin{bmatrix} 1 & \gamma_{12}(\omega) & \cdots & \gamma_{1m}(\omega) \\ \gamma_{21}(\omega) & 1 & \cdots & \gamma_{2m}(\omega) \\ \ldots & \ldots & \ldots & \ldots \\ \gamma_{m1}(\omega) & \gamma_{m2}(\omega) & \cdots & 1 \end{bmatrix} \tag{3.18}$$

Apparently, when $S(\omega)$ is determined, the spatial variation can be delineated by the coherence function matrix completely.

Obviously, the coherence function is also a complex function. Using the amplitude and phase angle expression, there is

$$\gamma_{kj}(\omega) = \left|\gamma_{kj}(\omega)\right| \exp\left[i\theta_{kj}(\omega)\right] \tag{3.19}$$

The amplitude $\left|\gamma_{kj}(\omega)\right|$ of the coherence function is called the *lagged coherence function*, and generally exists

$$\left|\gamma_{kj}(\omega)\right| \leq 1 \tag{3.20}$$

The phase angle $\theta_{kj}(\omega)$ is related to the propagation speed of the harmonic wave ω and the distance between two points (Oliveira and Hao, 1991)

$$\theta_{kj}(\omega) = \frac{\omega d_{kj}^{\mathrm{L}}}{v_a} \tag{3.21}$$

where d_{kj}^{L} is the projection of d_{kj}, the link between points k and j along the wave propagation direction, v_a is the apparent wave velocity of the ground motion. Setting a certain point as the starting point of the time coordinate, and denoting the time instants when the earthquake waves arrive at the points k and j as t_k and t_j, there is:

$$\frac{d_{kj}^{L}}{v_a} = t_k - t_j \tag{3.22}$$

Therefore, Eq. (3.19) can also be written as follows:

$$\gamma_{kj}(\omega) = \left|\gamma_{kj}(\omega)\right| exp\left[i\omega\left(t_k - t_j\right)\right] \tag{3.23}$$

The preceding deduction demonstrates that $\exp\left[i\theta_{kj}(\omega)\right]$ represents the difference of the arrival time of ground motion at two points, which is called the traveling wave effect, thus $\exp\left[i\theta_{kj}(\omega)\right]$ is generally called the traveling wave effect factor, whereas $\gamma_{kj}(\omega)$ reflects the coherence effect of the ground motions between the two points.

When the record of dense seismic array is used to study the ground motion, the absolute value of the coherence function, i.e., the lagged coherence function $\left|\gamma_{kj}(\omega)\right|$, is usually used to express the random field of the ground motion. It is noted that $\left|\gamma_{kj}(\omega)\right|$ generally reflects the influence of the phase angles. Follows are some typical examples.

(1) **Feng-Hu model** (Feng and Hu 1981)

Through analysis using the strong earthquake observed data from the Haicheng earthquake, 1975 in China, and the Niigata earthquake, 1964 in Japan, the following formula was proposed:

$$\gamma\left|\left(\omega, d_{kj}\right)\right| = \exp\left[-(\rho_1\omega + \rho_2)d_{kj}\right] \tag{3.24}$$

where ρ_1 and ρ_2 are the coherence parameters, of which the identified values from the Haicheng and Niigata earthquakes are respectively given by

$$\text{Haicheng earthquake}: \quad \rho_1 = 2\times 10^{-5} s/m, \rho_2 = 88\times 10^{-4} s/m;$$

$$\text{Niigata earthquake}: \quad \rho_1 = 4\times 10^{-4} s/m, \rho_2 = 19\times 10^{-4} s/m.$$

(2) **Loh-Yeh model** (Loh and Yeh 1988)

Through modeling with the observed data from the SMART-1 earthquake observation stations, the following formula was obtained:

$$\gamma\left|\left(\omega, d_{kj}\right)\right| = \exp\left(-\alpha\frac{\omega\, d_{kj}}{2\pi\, v_a}\right) \tag{3.25}$$

where α is the wavenumber of the ground motion, and from the existing 40 acceleration records, $\alpha = 0.125$ is identified (Loh 1991).

(3) **Qu-Wang model** (Qu et al. 1996)

Through modeling with the observed data from four earthquake observation stations including SMART-1, the following lagged coherence function was suggested:

$$\gamma\left|\left(\omega, d_{kj}\right)\right| = \exp\left(-a(\omega)d_{kj}^{b(\omega)}\right) \tag{3.26}$$

where

$$a(\omega) = \left(12.19 + 0.17\omega^2\right) \times 10^{-4} \tag{3.27}$$

$$b(\omega) = (76.74 - 0.55\omega) \times 10^{-2} \tag{3.28}$$

Figure 3.4 shows the comparison of the preceding three models as $\omega = 10\pi$. Obviously, the coherence function is totally different for a different statistical background.

If the auto-power spectral densities of all points are same, the power spectral density matrix of the ground motion random fields can be simplified as follows:

$$\mathbf{S}_B(\omega) = \mathbf{G}^*\mathbf{RG} \cdot S(\omega) \tag{3.29}$$

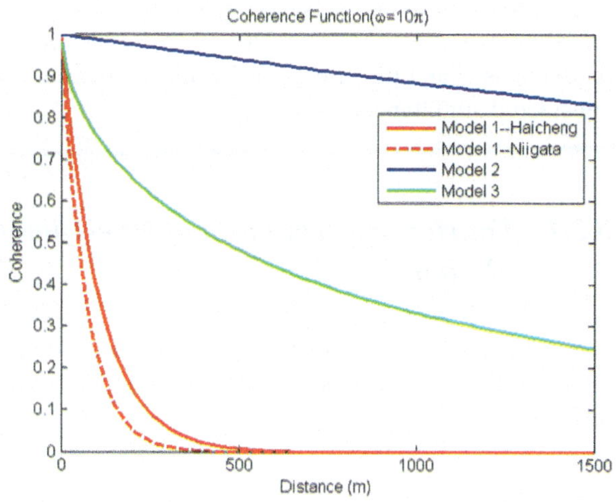

Fig. 3.4 Comparison of lagged coherence functions with different models

where $S(\omega)$ is the power spectral density at each point, **G** is a diagonal matrix representing the phase angle change of each point compared with the reference points

$$\mathbf{G} = \mathrm{diag}\left[e^{i\omega t_1}, e^{i\omega t_2}, \cdots, e^{i\omega t_m}\right] \tag{3.30}$$

and **R** is the matrix of the lagged coherence function

$$\mathbf{R} = \begin{bmatrix} 1 & |\gamma_{12}| & \cdots & |\gamma_{1m}| \\ |\gamma_{12}| & 1 & \cdots & |\gamma_{2m}| \\ \cdots & \cdots & \cdots & \cdots \\ |\gamma_{m1}| & |\gamma_{m2}| & \cdots & 1 \end{bmatrix} \tag{3.31}$$

3.3 Physically-Based Model

It is very difficult to model the non-stationary stochastic process or non-homogeneous random field based on the previous classical correlation function method or power spectral density method. In fact, only for the stochastic process with Gaussian normal nature, all the statistical features of the stochastic process could be obtained by the previous modeling methods. Unfortunately, for most of the practical cases, the Gaussian nature cannot be fully verified.

In the background, the ground motion is mainly affected by the earthquake fracture fault, the seismic wave propagation path, the site conditions and some other factors. Because most of these factors are beyond human control, notable random nature often appears in the observed ground motion. Considering the physical mechanism of ground motion, the physical random function model can be constructed (Li and Ai 2006; Wang and Li 2011, 2012). In this model, the random Fourier spectrum model is used and the effects of seismic source, propagation path and local site are considered simultaneously.

3.3.1 Fourier Spectral Form of One-Dimensional Ground Motion

According to the analysis of wave propagation equation in general linear homogeneous medium, the displacement of ground motion can be written as follows (Wang and Li 2011):

$$\begin{aligned} u(t) = \frac{1}{2\pi}\int_{-\infty}^{+\infty} & A_s(\alpha_1, \ldots, \alpha_s, \omega) \cdot H_{Ap}(\beta_1, \ldots, \beta_h, \omega, x) \cdot H_{As}(\gamma_1, \ldots, \gamma_l, \omega) \\ & \cdot \cos\left[\omega t + \Phi_s(\alpha_1, \ldots, \alpha_s, \omega) \cdot H_{\Phi p}(\beta_1, \ldots, \beta_h, \omega, x) \cdot H_{\Phi s}(\gamma_1, \ldots, \gamma_l, \omega)\right] d\omega \end{aligned} \tag{3.32}$$

where $A_s(\alpha_1, \ldots, \alpha_s, \omega)$ and $\Phi_s(\alpha_1, \ldots, \alpha_s, \omega)$ are the displacement amplitude and phase spectrum of seismic source, respectively; $H_{Ap}(\beta_1, \ldots, \beta_h, \omega, x)$ and $H_{\Phi p}(\beta_1, \ldots, \beta_h, \omega, x)$ are the amplitude and phase transfer function of seismic wave propagation path, respectively; $H_{As}(\gamma_1, \ldots, \gamma_l, \omega)$ and $H_{\Phi s}(\gamma_l, \ldots, \gamma_l, \omega)$ are the amplitude and phase transfer function of local site, respectively. α_i, β_i and γ_i are the physical parameters associated with seismic source, propagation path and local site, respectively. It needs to be emphasized here that, considering that the geometric scale of local site is much less than the scale of propagation path, both H_{As} and $H_{\Phi s}$ are independent of x. x is the distance from the local site to the seismic source.

The acceleration of ground motion corresponding to Eq. (3.32) is:

$$a(t) = \ddot{u}(t) = -\frac{1}{2\pi}\int_{-\infty}^{+\infty} \omega^2 A_s(\alpha_1, \ldots, \alpha_s, \omega) \cdot H_{Ap}(\beta_1, \ldots, \beta_h, \omega, x) \cdot H_{As}(\gamma_1, \ldots, \gamma_l, \omega)$$
$$\cdot \cos\left[\omega t + \Phi_s(\alpha_1, \ldots, \alpha_s, \omega) \cdot H_{\Phi p}(\beta_1, \ldots, \beta_h, \omega, x) \cdot H_{\Phi s}(\gamma_1, \ldots, \gamma_l, \omega)\right]d\omega \tag{3.33}$$

Obviously, the expression above is the Fourier spectral form of the acceleration time history which indicates the deterministic physical law governing the stochastic ground motion. Considering all the physical parameters are random variables, the general form of physical random function model of ground motion is:

$$a_R(t) = -\frac{1}{2\pi}\int_{-\infty}^{+\infty} A_s(\zeta_\alpha, \omega) \cdot H_{Ap}\left(\zeta_\beta, \omega, R\right) \cdot H_{As}\left(\zeta_\gamma, \omega\right)$$
$$\cdot \cos\left[\omega t + \Phi_s(\zeta_\alpha, \omega) \cdot H_{\Phi p}\left(\zeta_\beta, \omega, R\right) \cdot H_{\Phi s}\left(\zeta_\gamma, \omega\right)\right]d\omega \tag{3.34}$$

where $\zeta_\alpha = \left(\zeta_{\alpha_1}, \ldots, \zeta_{\alpha_s}\right)$, $\zeta_\beta = \left(\zeta_{\beta_1}, \ldots, \zeta_{\beta_h}\right)$ and $\zeta_\gamma = \left(\zeta_{\gamma_1}, \ldots, \zeta_{\gamma_l}\right)$ are the random vectors of the seismic source, propagation path and local site, respectively. ζ_{α_i}, ζ_{β_i}, ζ_{γ_i} are the random variables corresponding to parameters α_i, β_i and γ_i, respectively. R is the distance between the seismic source and the local site.

3.3.2 Seismic Source Spectrum

Seismic source models in seismology are mainly classified into kinematic models and dynamic models. The kinematic model is widely used in the earthquake engineering community, in which the Brune's source model is a general model which has few parameters and clear physical background [Brune 1970]. In this model, the fault surface is assumed as circular and the dislocation distributes uniformly on the fault surface. The rupture occurs instantaneously along with the circular. The shear stress wave caused by the shear stress drop propagates perpendicular to the dislocation surface. Figure 3.5 shows the illustration of Brune's dislocation model.

Considering the effects of the edges of the dislocation surface to the seismic wave propagation, the displacement near the fault is Brune (1970).

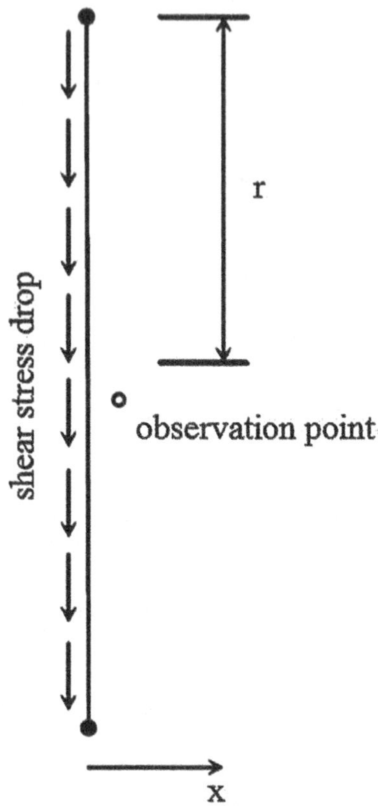

Fig. 3.5 The Brune's dislocation source model

$$u(x,t)|_{x\to 0} = \begin{cases} 0 & t \le 0 \\ (\sigma/\mu)\cdot \beta\tau\left(1 - e^{-(t/\tau)}\right) & t > 0 \end{cases} \tag{3.35}$$

in which σ is the shear stress drop, μ is the rigidity, β is the shear wave velocity, τ is a Brune's seismic source parameter.

Considering the physical parameters' randomness, the source amplitude spectrum $A_s(\omega)$ and phase spectrum $\Phi_s(\omega)$ are calculated and simplified from the Fourier transform of (3.35) as follows:

$$A_s(\xi_\alpha, \omega) = \frac{A_0}{\omega\sqrt{\omega^2 + \left(\frac{1}{\tau}\right)^2}} \tag{3.36}$$

$$\Phi_s(\xi_\alpha, \omega) = \arctan\left(\frac{1}{\tau\omega}\right) \tag{3.37}$$

where $\xi_\alpha = (A_0, \tau)$ is the random vector of the source physical model and $A_0 = \sigma\beta/\mu$ is a comprehensive random parameter.

3.3.3 Transfer Function of the Path

Physical factors influencing the amplitudes and phases of seismic waves propagating in the earth media are mainly the geometric spreading, reflection and refraction at the surfaces between the layers and the attenuation caused by internal friction in the media. A general amplitude spectrum is given by Aki and Richard (1980):

$$A(\omega) = A_0(\omega) \cdot \exp\left[-\frac{\omega x}{2cQ(\omega)}\right] \tag{3.38}$$

where $Q(\omega)$ is a dimensionless quantity which reflects the gross effect of the internal friction in the media, c is the average group velocity of the seismic P or S-waves.

Therefore, the transfer function of path is:

$$H(\omega) = \frac{A(\omega)}{A_0(\omega)} = \exp\left[-\frac{\omega x}{2cQ(\omega)}\right] = \exp[-K\omega x] \tag{3.39}$$

in which K is a parameter denoting the friction attenuation effect.

The observations indicate that c and Q are approximately independent of ω. As a result, K is a constant whose value takes 10^{-5} s/km that is calculated from the empirical average values of c and Q in the rock.

The phase form of the general seismic wave can be written as follows (Liao, 2002):

$$H_{\Phi p}(\omega, x) = -k(\omega) \cdot x \tag{3.40}$$

where $k(\omega)$ denotes the wavenumber-frequency relationship, by which the group velocity-frequency relationship can be calculated as follows:

$$c(\omega) = \frac{d\omega}{dk(\omega)} \tag{3.41}$$

The phase variation of seismic waves in the propagation caused by the reflection and refraction at the layer surfaces is hardly described by a general expression. In this book, we suggest to use an equivalent wavenumber-frequency formula such as Wong and Trifunac (1979):

$$k(\omega) = d \cdot \ln\left[(a + 0.5)\omega + b + \frac{1}{4c}\sin(2c\omega)\right] \tag{3.42}$$

Then the group velocity-frequency relationship corresponding to (3.42) is

$$c(\omega) = \frac{1}{\frac{dk(\omega)}{d\omega}} = \frac{(a + 0.5)\omega + b + \frac{1}{4c}\sin(2c\omega)}{d \cdot \left(a + cos^2(c\omega)\right)} \tag{3.43}$$

where *a*, *b*, *c* and *d* are empirical parameters and their values are determined from the realistic wavenumber-frequency relationship.

Then the Fourier amplitude and phase spectral transfer function of the path are denoted by:

$$H_{Ap}(\omega, R) = exp(KR\omega) \tag{3.44}$$

$$H_{\Phi p}\left(\xi_{\beta}, \omega, R\right) = -Rd \cdot \ln\left[(a+0.5)\omega + b + \frac{1}{4c}\sin(2c\omega)\right] \tag{3.45}$$

where *K* is a constant and *a*, *b*, *c* and *d* are random variables.

3.3.4 Local Site Effect

The seismic disasters indicate that the local engineering sites have obvious filter effect to the seismic waves from the bedrock. Because local site effects are largely independent of the propagation path effect, it is rational to separate the two effects in the ground motion modeling. Without loss of generality, an engineering site can be simulated as an equivalent single-degree-of-freedom system (Fig. 3.6) with the equation of motion

$$\ddot{x} + 2\zeta\omega_0\dot{x} + \omega_0^2 x = 2\zeta\dot{u}_g + \omega_0^2 u_g \tag{3.46}$$

where $\ddot{x}$, $\dot{x}$ and x respectively, represent the absolute acceleration, the absolute velocity and the absolute displacement of a point in the fixed coordinate system, while ω_0 and ξ are, respectively, the frequency and damping ratio of the site, $\dot{u}_g$ and u_g are, respectively, the velocity and the displacement of the input seismic wave at the bedrock.

Performing Fourier transform on both sides of the Eq. (3.46) and noticing the relationship between displacement and acceleration, the Fourier transform of the absolute acceleration is given as follows:

$$\ddot{X}(\omega) = \frac{\omega_0^2 + i2\zeta\omega_0\omega}{\omega_0^2 - \omega^2 + i2\zeta\omega_0\omega}\ddot{U}_g(\omega) \tag{3.47}$$

where $\ddot{U}_g(\omega)$ is the Fourier transform of the acceleration of the input seismic waves.

Considering the randomness of the equivalent damping ratio and predominant circular frequency, the amplitude spectral transfer function of the local site is

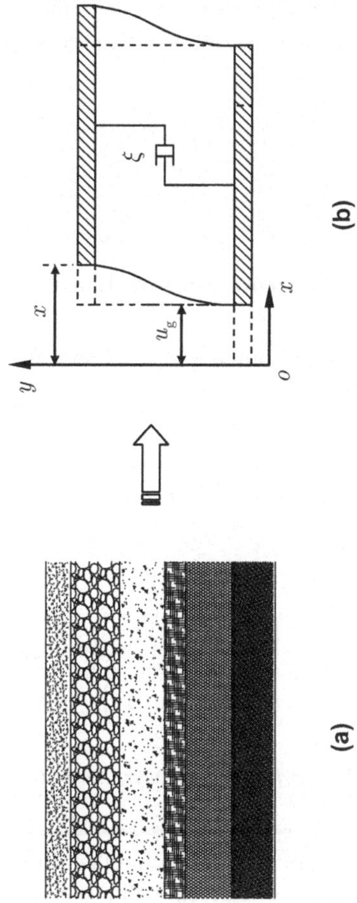

Fig. 3.6 Equivalent single-degree-of-freedom system

$$H_{As}(\xi_\gamma, \omega) = \left[\frac{1 + 4\xi^2 \left(\frac{\omega}{\omega_0}\right)^2}{\left(1 - \frac{\omega^2}{\omega_0^2}\right)^2 + 4\xi^2 \left(\frac{\omega}{\omega_0}\right)^2} \right]^{\frac{1}{2}} \tag{3.48}$$

where $\xi_\gamma = (\zeta, \omega_0)$ is the random vector of the local site model, ζ and ω_0 are both random variables.

Since the geometric scale of the local site is smaller far than that of the propagation path, it is suitable to neglect the phase variation when the seismic wave propagates through the local site. That is:

$$H_{\Phi s}(\xi_\gamma, \omega) = 0 \tag{3.49}$$

3.3.5 One-Dimensional Ground Motion Model

Combining all the effects of seismic source, propagation path and local site models, a physical random function model for the one-dimensional ground motions can be described as follows:

$$a_R(\xi, t) = \frac{1}{2\pi} \cdot \int_{-\infty}^{+\infty} A_R(\xi, \omega) \cdot \cos[\omega t + \Phi_R(\xi, \omega)] \cdot d\omega \tag{3.50}$$

where

$$A_R(\xi, \omega) = \frac{A_0 \omega \cdot e^{-K\omega R}}{\sqrt{\omega^2 + \left(\frac{1}{\tau}\right)^2}} \cdot \sqrt{\frac{1 + 4\xi_g^2 (\omega/\omega_g)^2}{\left[1 - (\omega/\omega_g)^2\right]^2 + 4\xi_g^2 (\omega/\omega_g)^2}} \tag{3.51}$$

$$\Phi_R(\xi, \omega) = \arctan\left(\frac{1}{\tau\omega}\right) - Rd \cdot \ln[(a + 0.5)\omega + b + \frac{1}{4c}\sin(2c\omega)] \tag{3.52}$$

where $\xi = (\xi_\alpha^T, \xi_\beta^T, \xi_\gamma^T)$ is a random vector which reflects all the random factors.

When ξ in $A_R(\xi, \omega)$ and $\Phi_R(\xi, \omega)$ is given with a specific realization value, the definite sample function will be presented. This provides the possibility of using the observed seismic records to identify the parameters of the model and therefore complete the modeling of ground motion. After the basic random variables are identified for every sample of an assemble of seismic records, the probability distribution function of the random vector ξ can be identified.

Table 3.1 gives some typical probability distributions of parameters, where μ and σ denote mean and standard deviation of the natural logarithm for lognormal distribution, k and θ are shape parameter and scale parameter, respectively. Figure 3.7

Table 3.1 Probability distribution of random parameters

Random variable	Probability distribution	Parameters
A_0 (g·s/rad)	lognormal	$(\mu,\sigma) = (-1.27,0.83)$
τ (s/rad)	lognormal	$(\mu,\sigma) = (-1.24,1.34)$
ξ_g (/rad)	gamma	$(k,1/\theta) = (5.13, 0.08)$
ω_g	gamma	$(k,1/\theta) = (2.24, 7.41)$

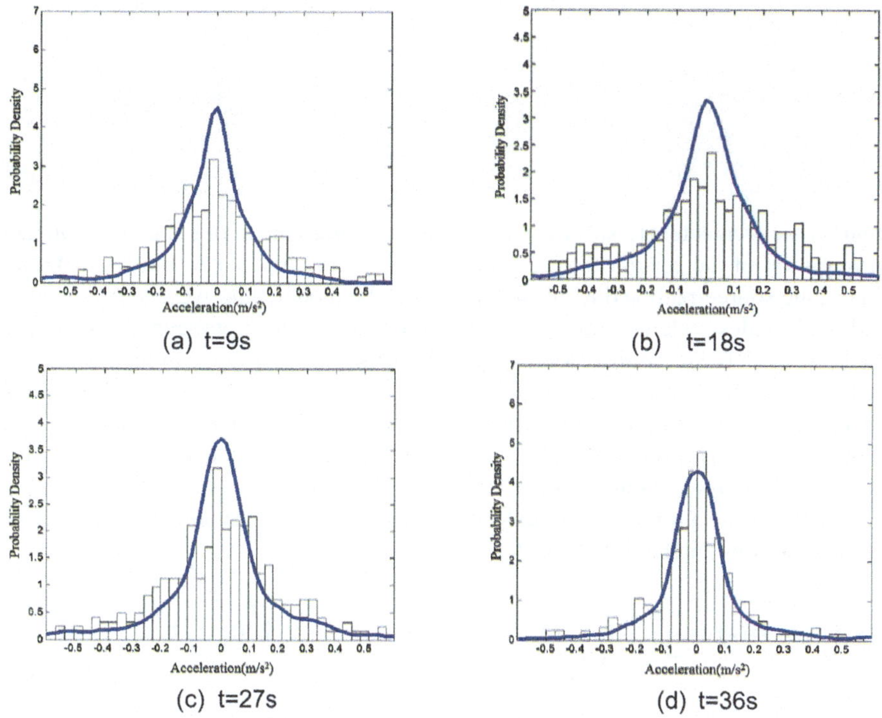

Fig. 3.7 Comparison of the statistical histograms of recorded seismic ground motions and probability distribution functions presented by physical model

gives the statistical histograms and corresponding probability distribution function of these parameters (Wang and Li, 2012; Li and Song, 2013).

3.3.6 Physical Random Field Model of Ground Motions

As mentioned in Sect. 3.1, the basic model of ground motion should be modeled as a spatial random field. The essential physical factors, which affect the seismic wave

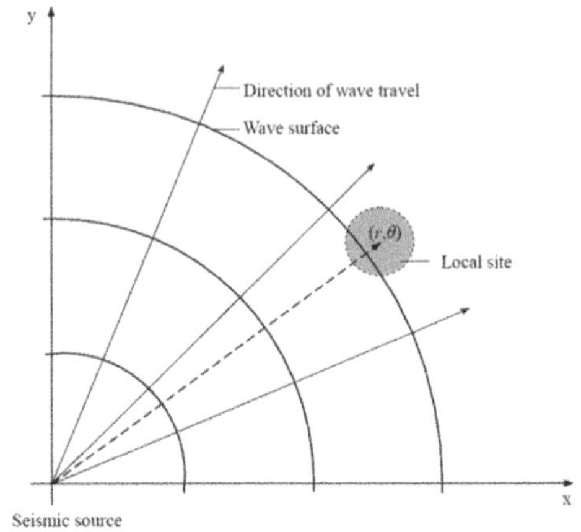

Fig. 3.8 Spherical seismic wave field on bedrock

field on the local site, are still the source-path-site mechanism of the generation and propagation of the seismic waves. Considering the seismic waveforms in different scales, the source-path and local site effects can be modeled in two scales, i.e. large and small scales. In large scale, the seismic point source and homogeneous isotropic medium models are applied to calculate the seismic wave field on the surface of bedrock; In small scale, the soft solid cover is simplified as a filter and the local coordinate system on the engineering site is introduced. A polar coordinate system shown in Fig. 3.8 is proposed to describe the two-dimensional spherical seismic wave field on the bedrock (Wang and Li 2012).

According to a similar physical analysis, the physical random field model of ground motions can be derived as follows:

$$a_R(\xi, \eta, r, r_l, t) = \frac{1}{2\pi} \cdot \int_{-\infty}^{+\infty} A_R(\xi, \eta, \omega, r, r_l) \cdot \cos[\omega t + \Phi_R(\xi, \eta, \omega, r, r_l)] \cdot d\omega \tag{3.53}$$

where

$$A(\xi, \eta, \omega, r, r_l) = \frac{A_0 \omega \cdot e^{-K\omega r}}{\sqrt{\omega^2 + \left(\frac{1}{\tau}\right)^2}} \cdot \sqrt{\frac{1 + 4\xi_g^2 \left(\frac{\omega}{\omega_g}\right)^2}{\left[1 - \left(\frac{\omega}{\omega_g}\right)^2\right]^2 + 4\xi_g^2 \left(\frac{\omega}{\omega_g}\right)^2}} \cdot e^{-\frac{\alpha_0 \omega r_1}{2}} \tag{3.54}$$

$$\Phi(\xi, \eta, \omega, r, r_l) = \arctan\left(\frac{1}{\tau\omega}\right) - r \cdot d \cdot \ln\left[(a + 0.5)\omega + b + \frac{1}{4c}\sin(2c\omega)\right] - r_l \cdot \frac{\omega}{c_g} \tag{3.55}$$

Probability density functions of α_0 and c_g are

$$f(\alpha_0) = \frac{1}{\sqrt{2\pi}\sigma_{\alpha_0}} \cdot e^{-\frac{\left(\alpha_0 - \mu_{\alpha_0}\right)^2}{2\sigma_{\alpha_0}^2}} \tag{3.56}$$

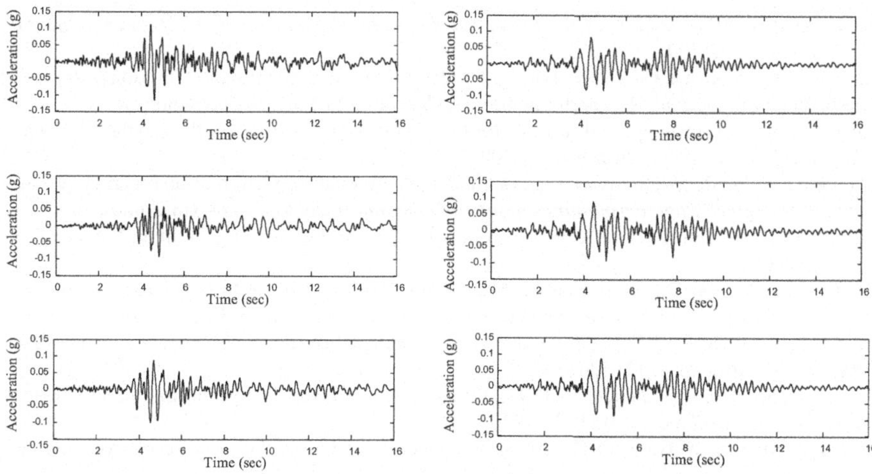

Fig. 3.9 Real seismic records (left side) and theoretical simulations (right side)

$$f(c_g) = \frac{1}{\sqrt{2\pi}\, c_g \sigma_{c_g}} \cdot e^{-\frac{(\ln c_g - \mu_{cg})^2}{2\sigma_{cg}^2}} \tag{3.57}$$

where $\eta = (\alpha_0, c_g)$ is a random vector.

Using seismic records, the statistical parameter in the above model can be identified. Figure 3.9 is a set of comparing results of the real seismic records and the theoretical simulations using the proposed model.

References

Aki K, Richards P G. (1980). Quantitative seismology theory and methods. San Francisco: W. H. Freeman and Company.

Amin, M., & Ang, A. H. S. (1968). Nonstationary stochastic model of earthquake motion. *Engineering Mechanics Division, ASCE, 94*(EM2), 559–584.

Brune, J. N. (1970). Tectonic stress and the spectra of seismic shear waves from earthquakes. *Journal of Geophysical Research, 75*(26), 4997–5009.

Burg, J. P. (1967), Maximum entropy spectral analysis, presented at the 37th Meeting of the Society of Exploration Geophysicists; reprinted in D. G. Childers, ed. (1978), *Modern Spectrum Analysis, IEEE Press*, pp. 34–41.

Clough, R. W., & Penzien, J. (1993). *Dynamics of structures*. New York: McGraw Hill.

Feng, Q. M., & Hu, Y. X. (1981). Spatial correlation model of ground motion. *Earthquake Engineering and Engineering Vibration, 1*(2), 1–8.

Hu, Y. X., Zhou, X. Y. (1962). A review on statistical theory for seismic forces. Collected Research Report on. Earthquake Engineering, pp. 21–32.

Huang, N. E., Shen, Z., Long, S. R., Wu, M. C., Shih, H. H., & Zheng, Q., et al. (1998). The empirical mode decomposition and the hilbert spectrum for nonlinear and non-stationary time

series analysis. *Proceedings of the Royal Society A: Mathematical, Physical and Engineering Sciences*, 454, 903–995.
Jennings P. C., Housner G. W., & Tsai N. C. (1968). Simulated earthquake motions. Report of Earthquake Engineering Research Laboratory, California Institute of Technology.
Kanai, K. (1957). Semi-empirical formula for the seismic characteristics of the ground. *Bulletin of the Earthquake Research Institute, 35,* 309–325.
Li, J., & Ai, X. Q. (2006). Study on random model of earthquake ground motion based on physical price. *Earthquake Engineering and Engineering Vibration, 26*(5), 21–26. (in Chinese).
Liao, Z. P. (2002), Introduction to Wave Motion Theories in Engineering. Beijing: Science Press (in Chinese).
Loh, C. H., & Yeh, Y. T. (1988). Spatial variation and stochastic modelling of seismic differential ground movement. *Earthquake Engineering Structural Dynamics, 16*(4), 583–596.
Loh, C. H., & Lin, S. G. (1990). Directionality and simulation in spatial variation of seismic waves. *Engineering Structures*, 12: 134–143.
Oliveira, C. S., Hao, H., Penzien, J. (1991). Ground motion modeling for multiple-input structural analysis. *Structural Safety*, 10(1–3): 79–93.
Qu, T. J., Wang, J. J., & Wang, Q. X. (1996). A practical model of spatial variation of ground motion power spectrum. *Acta Seismologica Sinica, 18*(1), 55–62.
Ruiz, P., & Penzien, J. (1969). *Artificial generation of earthquake accelerograms.* Berkeley: Report of University of California.
Tajimi, H. (1960). A statistical method of determining the maximum response of a building structure during an earthquake. Proc 2nd WCEE. *Tokyo, II,* 781–798.
Vanmarcke, E. (1983). Random Fields. Cambridge Ma Mit Press.
Wang, D., & Li, J. (2011). Physical random function model of ground motions for engineering purposes. *Science China: Technological Sciences, 54*(1), 175–182.
Wang, D., & Li, J. (2012). A random physical model of seismic ground motion field on local engineering site. *Science China: Technological Sciences, 55*(7), 2057–2065.
Wong, H. L., & Trifunac, M. D. (1979). Generation of artificial strong motion accelerograms. *Earthquake Engineering and Structural Dynamics, 7*(6), 509–527.

Chapter 4
Seismic Performance Evaluation of Buried Pipelines

4.1 Seismic Damage of Buried Pipelines

4.1.1 Pipeline Damage in Past Earthquakes

Buried pipelines are the major component of many lifeline engineering systems, such as water supply systems, gas supply systems and heating supply systems. During past earthquakes, buried pipelines suffered serious damages and some of them cause serious secondary disasters.

Seismic investigation on buried pipelines originated from the early of 1920's. After the Kanto earthquake, Japan, in 1923, the buried pipelines in Tokyo were investigated in detail. The investigation indicated that over 80% joints, were damaged, about 71% of which were leaky. Afterward, the damages of lifeline engineering systems (especially for the buried pipelines) during earthquakes which is close to modern cities are investigated in detail. For example, after the San Fernando earthquake, American in 1971, the buried pipelines of Los Angeles city were investigated widely (Wang, Sun, & Shen, 1985). The investigation indicated that 856 breaks appeared on water pipelines and the average damage ratio of water cast iron pipelines was 0.95/km in the area where the seismic intensity was between VIII and IX. Meanwhile, 450 breaks appeared on gas pipelines of which were mostly 50–100 mm welded steel pipes and the damage ratio was 0.12/km. On July 28, 1976, a strong earthquake with a magnitude of 7.8 attacked Tangshan, China. The seismic intensity of epicenter was as high as XI. Beijing and Tianjin, the capital and an important municipality of China, were also influenced by this strong earthquake. The earthquake seismic of Tianjin was VII–IX. Table 4.1 illustrates the seismic investigation of water supply pipelines in different areas with different site conditions (Liu, 1986). In 1995, a strong earthquake with a magnitude of 7.2 happened in Kobe, Japan. Like other strong earthquakes, the buried pipelines suffered serious damages. Based on these historic investigations, in 2001, the American Lifeline Engineering Associate summarized the earthquake damages of buried pipelines systemically (ALA, 2001).

J. Li and W. Liu, *Lifeline Engineering Systems*,
https://doi.org/10.1007/978-981-15-9101-3_4

Table 4.1 Damage ratio of buried pipelines in different areas subjected to the Tangshan earthquake

Area	Seismic Intensity	Soil classification	Pipe diameter (mm)	Pipe length (km)	Number of damages	Average damage ratio (/km)
Tianjin	VII–VIII	III	75–1000	870.0	161	0.18
Tanggu	VIII	III	75–600	79.50	332	4.18
Hangu	IX	III	–	–	–	10.00
Tangshan	X–XI	II	75–600	111.00	446	4.00

4.1.2 Damage Characteristics of Buried Pipelines

The major effects of earthquake on buried pipelines are permanent ground deformation and ground motion. The former includes fault rupture, soil liquefaction, landslip and surface collapse. In the area where permanent ground deformation covers, the damage ratio of buried pipelines increases significantly. The latter means the seismic wave propagation, which will cause excessive deformation of buried pipelines and lead to the damage. The existing investigations indicate that most damages of buried pipelines attribute to seismic wave propagation. When the site consists of non-homogeneous or soft soil layer, the pipeline damages increase significantly (Ariman & Muleski, 1981). Moreover, when the ground peak acceleration is the same, pipeline strain caused by the surface wave is larger than that caused by the body wave (Sato et. al., 1988). Generally, the damages caused by permanent ground deformation are impossible to be avoided. Thus, some specified seismic measures should be adopted when the pipelines pass through these areas.

The damages of straight buried pipeline can be classified into three types.

(1) Pipeline joint damage. For segmented pipeline, the damages include the seal material loosening and joint pulled out. For continuous steel pipeline, the damages include the weld crack and bolt loosening.
(2) Pipeline body damage. It includes longitudinal or oblique cracks and pipeline rupture.
(3) Junction damage. It means the damages of tees, valves and junctions between pipelines and buried structures.

Among the three damage types, pipeline joint damage is the most common. For example, in the Tokachi earthquake, Japan, the joint damages occupied 75% of all pipeline damages. In the Tangshan earthquake, China, the joint damages reached 79% of all pipeline damages. Nevertheless, in the area with soil liquefaction or fault, the second type of damages is also significant.

4.1.3 *Factors Affecting Buried Pipeline Damages*

The most influential factors for buried pipeline damages are seismic intensity, site, pipeline material, pipeline diameter and the formulation of pipeline joints.

Many seismic surveys indicate that the damage ratio of buried pipeline increases with the increase of seismic intensity. However, site condition has a significant influence on this general tendency. The investigation of the Tangshan earthquake indicates that the pipeline damage ratio in soft-soil site under low seismic intensity may be larger than that in hard-soil site (Table 4.1). This phenomenon could also be found in many other events of strong earthquakes. Generally, the damage ratio of buried pipeline is small in hard-soil site, while large in the soft-soil site. Figure 4.1 shows the pipeline damage in different sites in the Urakawaoki earthquake, Japan, in 1982 (Hou, 1990).

The materials of buried pipelines could be steel, cast iron, asbestos cement, plastic and prestressed concrete. Table 4.2 lists the damages of pipelines with different materials during typical strong earthquakes (Sun, 1986). It is shown that the damage ratio of steel pipelines varies significantly because the performance of the pipeline is affected by the corrosion. The performance of prestressed concrete pipelines is the best and that of cast iron pipeline is better than that of asbestos cement pipelines. The seismic performance of plastic pipelines, especially the high-strength PVC pipelines developed recently, is uncertain because it has seldomly been tested by actual earthquakes.

Fig. 4.1 Pipeline damage in different sites

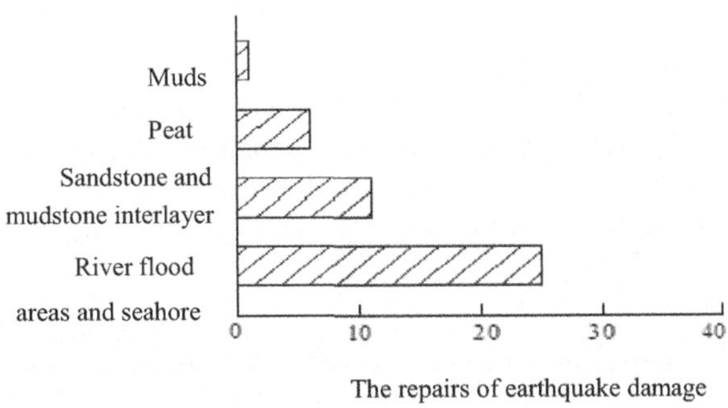

Table 4.2 Damage ratios of pipelines with different materials (/km)

Region material	Yingkou city (VIII)	Yingkou county (IX)	Panshan (IX)	Sendai (IX)
Steel	11.4	2.10	0.7	0.01
Cast iron	1.06	0.96	1.6	0.05
Asbestos cement	2.00	3.00	1.3	0.81
Plastic	–	–	–	0.31
Prestressed concrete	0	–	–	–

Pipeline stiffness can reduce the soil deformation around it. Thus, the damage ratio of pipelines with large diameter is relatively low compared with the pipeline with small diameter.

Comparing with the pipeline body, the pipeline joint is a weak part under earthquake. Generally, pipeline junction can be divided into rigid junction and flexible junction. Rigid junction includes weld, threaded connection, lead, auto-stressed cement and asbestos cement joints. Flexible junction includes rubber bell-and-spigot joint and bolt-flange junction. The damage ratio of flexible junction is much lower than that of rigid junction as the limit deformation of flexible junction is large.

In addition to the above major factors, the service time, design and construction quality are also important factors. However, the available data related to these factors is few.

4.1.4 Empirical Statistics of Damage Ratio

Using available historic seismic records, the empirical function relationship between pipeline damage ratio and influential factors can be established. The empirical statistics model can be used to estimate pipeline damage ratio or seismic reliability (Li, 1992).

Assume the pipeline damage follows Poisson distribution along with pipeline length, the following equation can be used to give pipeline failure probability.

$$P_f = 1 - e^{-\lambda L} \tag{4.1}$$

where λ(1/km) is the average damage ratio and L(km) is the pipeline length.

The key parameter of above equation is the average damage ratio λ. In the early research, it was expressed as a function of seismic intensity. For example, in the late 1990s, the following equation is used to calculate the average damage ratio (Hwang & Lin, 1998).

$$\lambda = C_d C_g 10^{0.8(MMI-9)} \tag{4.2}$$

where MMI is the modified Mercalli intensity, C_d is the influential factor of pipeline diameter and C_g is the influential factor of site soil:

$$C_{\mathrm{d}} = \begin{cases} 1.0 & D < 250\ \mathrm{mm} \\ 0.525 & 250\ \mathrm{mm} \le D < 500\ \mathrm{mm} \\ 0.25 & 500\ \mathrm{mm} \le D < 1000\ \mathrm{mm} \\ 0.01 & D \ge 1000\ \mathrm{mm} \end{cases} \tag{4.3}$$

where D is pipeline diameter.

$$C_g = \begin{cases} 0.5 & \text{Site soil I} \\ 1.0 & \text{Site soil II} \\ 2.0 & \text{Site soil III} \\ 5.0 & \text{Site soil IV} \end{cases} \tag{4.4}$$

Apparently, the two equations above do not consider the influence of pipeline materials and joints.

Existing research results indicate that seismic wave's peak velocity and peak displacement directly influence the seismic damage ratio of buried pipeline, while seismic intensity, such as *MMI* can only give a rough result (Li & Li, 1992). Moreover, the damage investigations of the earthquakes all over the world in recent 30 years has provided a number of data for the buried pipelines with different materials and different diameters. Based on this background, American Lifeline Alliance proposed the following method to compute the average seismic damage ratio for buried pipelines (ALA, 2001).

1. The seismic damage ratio caused by seismic wave propagation:

$$\lambda = 0.00475 \cdot K_1 \cdot PGV \tag{4.5}$$

where PGV is the seismic efficient peak velocity (cm/s) and K_1 is the adjustment factor considering all kinds of influential factors and its value is shown in Table 4.3.

2. The seismic damage ratio caused by ground permanent deformation

$$\lambda = 1.427 \cdot K_2 \cdot PGD^{0.32} \tag{4.6}$$

where PGD is the seismic efficient peak displacement (cm) and K_2 is the adjustment factor considering all kinds of influential factors and its value is shown in Tab. 4.4.

It is worth pointing out that the average damage ratio λ in Eq. (4.1) is difficult to define definitely due to the complex seismic damage state and inconsistent state division standard. Many research regard the damage from the pipeline leakage to the water supply stoppage as one damage state. In fact, even in ALA report, it is only pointed out that, among the damages caused by seismic wave propagation, approximately 15% damages are water supply stoppage and approximately 85% damages are pipeline leakage. Among the pipeline damages caused by permanent ground deformation, approximately 50% damages are water supply stoppage and approximately 50% damages are pipeline leakage (ALA, 2001). Apparently, the description is inexact. Therefore, the result given by empirical statistics cannot be applied to the accurate seismic reliability analysis.

Table 4.3 Adjustment factor K_1

Pipeline material	Joint	Site soil	Diameter (mm)	K_1
Cast iron	Cement	Any	100–300	1.0
Cast iron	Cement	Corrosive	100–300	1.4
Cast iron	Cement	Non-corrosive	100–300	0.7
Cast iron	Gasket	Any	100–300	0.8
Steel	Arc welding	Any	100–300	0.6
Steel	Arc welding	Corrosive	100–300	0.9
Steel	Arc welding	Non-corrosive	100–300	0.3
Steel	Arc welding	Any	>390	0.15
Steel	Gasket	Any	100–300	0.7
Steel	Bolt	Any	100–300	1.3
Steel	Riveting	Any	100–300	1.3
Asbestos cement	Gasket	Any	100–300	0.5
Asbestos cement	Cement	Any	100–300	1.0
Steel coating with concrete	Arc welding	Any	>390	0.7
Steel coating with concrete	Cement	Any	>390	1.0
Steel coating with concrete	Gasket	Any	>390	0.8
PVC	Gasket	Any	100–300	0.5
Ductile cast iron	Gasket	Any	100–300	0.5

Table 4.4 Adjustment factor K_2

Pipeline material	Joint	K_2
Ductile cast iron	Cement	1.0
Ductile cast iron	Gasket	0.8
Ductile cast iron	Machinery	0.7
Steel	Welding (no corrosion)	0.15
Steel	Gasket	0.7
Asbestos cement	Gasket	0.8
Asbestos cement	Cement	1.0
Steel coating with concrete	Welding	0.6
Steel coating with concrete	Cement	1.0
Steel coating with concrete	Gasket	0.7
PVC	Gasket	0.8
Ductile cast iron	Gasket	0.5

4.2 Seismic Response Analysis of Buried Pipelines

4.2.1 Pseudo-static Analysis Method

When seismic wave propagates, the soil deformations of different site points are different, which results in pipeline axial and bending deformations. Generally, for straight buried pipelines, the bending stress is small compared with the axial stress. Thus, the bending stress is usually neglected in seismic response analysis. For pipeline axial deformation, many researchers have found that the dynamic response for deformation computation of pipeline is small because the light pipeline weight and large damping of the soil around pipelines (Ariman & Muleski, 1981). As the result, the pseudo-static analysis method is enough for calculating the seismic deformation of buried pipelines.

Figure 4.2 shows an analytical model in which a series of distribution springs between the buried pipeline and the surrounding soil is used to simulate the pipe-soil interaction. Similarly, the segmented pipeline joints can be simulated by springs. Seismic displacement wave, u_g, is assumed to propagate along the pipeline axis with velocity c. With finite element discretization, a pseudo-static incremental equilibrium equation of buried pipelines is given as:

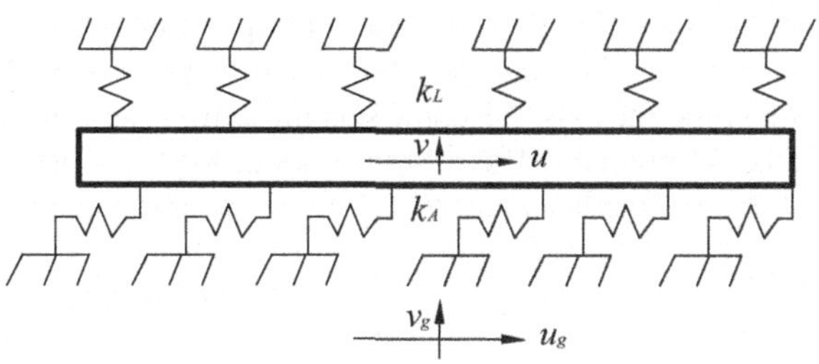

Fig. 4.2 Analytical model for pipelines

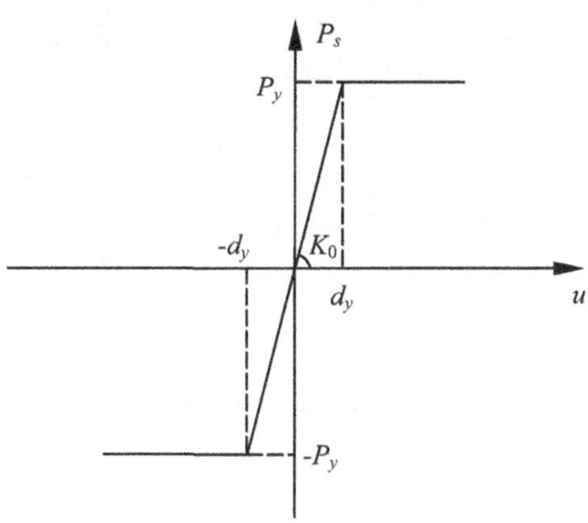

Fig. 4.3 Force-displacement relationship of soil spring

$$(\boldsymbol{K}_P + \boldsymbol{K}_s)\Delta\boldsymbol{u} = \boldsymbol{K}_s\Delta\boldsymbol{u}_g - \boldsymbol{K}_P\boldsymbol{U}_j - \boldsymbol{P}_{sj} \tag{4.7}$$

where $\boldsymbol{K}_P$ is the axial stiffness matrix of buried pipeline, $\boldsymbol{K}_s$ is the stiffness matrix of soil spring and the force-displacement relationship is shown in Fig. 4.3, $\Delta\boldsymbol{u}$ is the pipeline node displacement increment, $\Delta\boldsymbol{u}_g$ is the seismic wave displacement increment, $\boldsymbol{U}_j$ is the pipeline node displacement at the time of j and $\boldsymbol{P}_{sj}$ is the soil spring force at the time of j.

Considering the elastoplastic properties of soil springs and pipeline joints, the equation above is a non-linear equation. Thus, at each increment of analysis process, the stiffness matrix $\boldsymbol{K}_s$ and $\boldsymbol{K}_P$ should be modified. Usually, the soil spring stiffness takes the value of $0.6\pi D$, where D is the pipeline diameter. The soil spring yielding displacement takes a value of 0.2–0.5 cm for sand and 0.5–1.0 cm for clay.

The computational accuracy for the pseudo-static analysis presented here is determined by the seismic wave displacement input, the pipe-soil interaction relationship and the parameter values. However, these three factors are random in engineering practice. Thus, the feasible analysis should resort to an analytical approach for stochastic structures subjected to stochastic seismic waves.

For the seismic deformation of buried pipelines, some semi-empirical and semi-theoretical methods have been proposed based on seismic wave or sine wave.

(1) **Simplified method I**

It is assumed that the deformation of pipeline is the same as that of the surrounding soil. That is, the pipeline axial strain is equal to the strain of the surrounding soil along pipeline axis. Also, the seismic wave is assumed as a shear wave with unvaried shapes (Newmark, 1967; Shinozuka & Koike, 1979). According to the elastic wave theory, the displacement of a general wave propagating can be expressed as follows:

$$v = f(x + c_s t) \tag{4.8}$$

where v is the soil particle displacement and its direction is perpendicular to x axis. c_s is the shear wave velocity, t is the time and $f(\cdot)$ is a time history function.

It is noticed that the vibration velocity of soil particles is

$$V = \frac{\partial v}{\partial t} \tag{4.9}$$

And the soil strain is

$$\gamma = \frac{\partial v}{\partial x} \tag{4.10}$$

Substituting Eq. (4.8) in Eqs. (4.9) and (4.10), it can be deduced:

$$\gamma(t) = \frac{V(t)}{c_s} \tag{4.11}$$

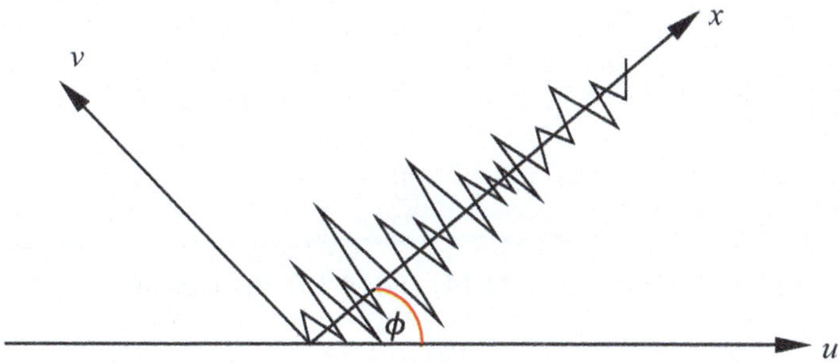

Fig. 4.4 Coordinates for seismic analysis of pipelines

Equation (4.11) indicates that the soil shear strain is proportional to the vibration velocity of soil particles and is inversely proportional to the shear wave velocity of the site.

It must be pointed out here that the shear wave velocity of the site is not equal to that of the surrounding soil. The reason is that seismic wave propagates in the soil within a certain depth. Thus, the shear velocities of soils on the surface and in a certain depth should be considered. A proposed shear velocity is given by (Han, 2002):

$$c_s = \frac{2c_1c_2}{c_1 + c_2} \tag{4.12}$$

where c_1 is the average shear wave velocity of soil on surface and c_2 is the shear wave velocity of soil in a certain depth, such as the bedrock.

For a shear wave, the direction of shear strain is perpendicular to the wave propagation direction. Assuming that the angle between the pipeline axis x' and the wave propagation direction is ϕ, the projection of shear deformation v on the pipeline axis is (see Fig. 4.4).

$$u = v \sin\phi = f(x + c_s t)\sin\phi \tag{4.13}$$

The relationship between x and x' is

$$x = x' \cos\phi$$

The soil particle vibration velocity is:

$$V' = \frac{\partial u}{\partial t} \tag{4.14}$$

and the axial strain is:

$$\varepsilon = \frac{\partial u}{\partial x'} \tag{4.15}$$

Substituting Eq (4.13) into Eqs. (4.14) and (4.15), there exists:

$$\varepsilon(t) = \frac{V(t)}{c_s}\sin\phi\cos\phi \tag{4.16}$$

The strain in Eq. (4.16) is also the soil stain alone pipeline axis. It is consistent with the assumption that the pipeline strain is equal to the surrounding soil strain, the pipeline axial strain is equal to ε. Considering Eq. (4.11), the pipeline axial strain can be stated as:

$$\varepsilon(t) = \frac{\gamma(t)}{2}\sin 2\phi \tag{4.17}$$

Apparently, ε takes the maximum value when $\phi = 45°$, and the maximum value is:

$$\varepsilon_{\max} = \frac{1}{2}\gamma_{\max} = \frac{V_{\max}}{2c_s} \tag{4.18}$$

However, in practice, the average pipeline stiffness is different from the soil stiffness even though considering the influence of pipeline joints. This difference will lead to the slippage between the pipeline and the surrounding soil, i.e. the pipeline deformation is different from the surrounding soil deformation. Considering the slippage, a transfer coefficient, ξ, is introduced. That is, the pipeline axial strain is denoted by:

$$\varepsilon_P(t) = \xi\varepsilon(t) \tag{4.19}$$

and

$$\varepsilon_{P\max} = \xi\varepsilon_{\max} \tag{4.20}$$

Theoretically, ξ should be derived using the elastic foundation beam model shown in Fig. 4.2. However, the value of ξ given by the model is overestimated when the diameter of straight pipeline is small, and underestimated when the diameter is large. Thus, some empirical equations based on the seismic damages records are adopted in engineering practice. For example, an empirical formula based on the data from the Haicheng and Tangshan earthquakes China is:

$$\xi = \frac{1}{1 + \frac{EAD}{2c_s^2}} \tag{4.21}$$

where E and A are the elastic modulus and the section area of pipeline, respectively, D is the pipeline average diameter and c_s is the average shear wave velocity of the site.

Moreover, the pipeline strain is usually amplified at the elbow, tee and cross of pipelines because of the stress concentration. Therefore, a magnification coefficient α should be introduced:

$$\varepsilon_P(t) = \alpha\xi\varepsilon(t) \tag{4.22}$$

and

$$\varepsilon_{P\max} = \alpha\xi\varepsilon_{\max} \tag{4.23}$$

Using finite element method, Liu and Hou (1990) gave a strain magnification coefficient as follows:

$$\alpha = \begin{cases} 1.5 & \text{Elbow} \\ 1.8 & \text{Tee} \\ 2.0 & \text{Cross pipe} \\ 2.5 & \text{One end fixed pipe} \end{cases} \tag{4.24}$$

Apparently, the deformation of a pipeline is

$$u(t) = \varepsilon(t) \cdot l \tag{4.25}$$

$$u_{\max} = \varepsilon_{P\max} \cdot l \tag{4.26}$$

where l is the length of the pipeline.

For the segmented pipelines, because the stiffness of pipeline body is much larger than that of joint, it is suitable to assume that the pipeline deformation concentrates on the pipeline joint, i.e. a pipeline joint deformation is equal to the pipeline deformation of a certain length.

For the continuous pipelines, its stress is

$$\sigma(t) = E\varepsilon_P(t) \tag{4.27}$$

$$\sigma_{\max} = E\varepsilon_{P\max} \tag{4.28}$$

The above seismic response analysis method is available for the arbitrary wave displacement function. Thus, it is suitable for general seismic wave input. When seismic wave velocity $V(t)$ is known, Eq. (4.16) or (4.19) can be used to calculate the pipeline axial strain and Eq. (4.25) can be used to calculate the pipeline deformation. When only the peak velocity of the seismic wave is known, Eq. (4.20) can be used to calculate the maximum pipeline deformation under a given exceedance probability. Therefore, simplified method I is a general and practical method.

(2) **Simplified method II**

Assuming that the seismic wave is a sine wave, a simplified method can be used to estimate the buried pipeline strain and displacement (Kuesel, 1969).

Let the soil displacement in the seismic wave plane be

$$Y_s = u_A \sin \frac{2\pi}{L} x \tag{4.29}$$

where u_A is the peak wave displacement, L is the seismic wave length and is approximately equal to the product of the shear wave velocity of the site soil and the predominant period of the site soil.

Just as Fig. 4.5 shows, when the angle between the seismic wave propagation direction x and the pipeline direction x' axis is ϕ, the soil displacement along the pipeline axis is

$$u_s = u_A \sin\phi \sin\frac{2\pi x'}{L'} \tag{4.30}$$

where L' is the projection of wave length on the pipeline axis and is called apparent wave length.

$$L' = \frac{L}{\cos\phi} \tag{4.31}$$

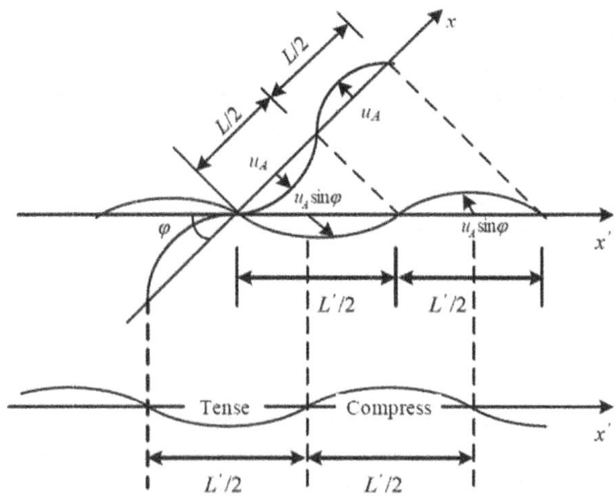

Fig. 4.5 Propagation of sine wave in the soil

Then the pipeline axial strain is:

$$\varepsilon = \frac{\partial u_s}{\partial x'} = \frac{\pi}{L} u_A \sin 2\phi \cos \frac{2\pi x'}{L'} \tag{4.32}$$

When $\phi = 45°$, maximal axial strain is

$$\varepsilon_{max} = \frac{\pi u_A}{L} \cos \frac{2\pi x'}{L'} \tag{4.33}$$

In Fig. 4.5, the axial deformation of pipeline caused by seismic wave is in tension within a half apparent wave length and in compression within the next half apparent wave length[1]. Thus, for a sine wave, the half apparent wave length can be regarded as a standard computation length, by which the maximal deformation is denoted by:

$$\Delta L = \frac{\pi u_A}{L} \int_{-\frac{L'}{4}}^{\frac{L'}{4}} \cos \frac{2\pi x'}{L'} dx' = \sqrt{2} u_A \tag{4.34}$$

Usually, the above deformation is resisted by the pipeline joints in the standard computation length. Similarly, considering the slippage, the axial deformation transfer coefficient ξ should be introduced. For the specified location, such as tee and elbow, the strain magnification coefficient should be introduced as well.

The peak displacement in Eq. (4.29) can be obtained from the result of seismic microzonation of the site. If only the peak acceleration in the result is available, the following equation can be used to calculate the peak displacement:

$$u_A = \frac{a \cdot T^2}{4\pi^2} \tag{4.35}$$

where a is the peak seismic acceleration, T is the predominant period of site soil.

It is worth pointing out that the maximal pipeline strain or displacement given by Eq. (4.33) or (4.34) is the maximum value with a specified seismic exceedance probability.

In view of that the above method cannot consider the real earthquake ground motion input, the simplified model I is generally recommended in the seismic response analysis of pipelines.

[1]It is noticed that it is only valid when the seismic wave is sine wave. For general seismic wave, it might not be true.

4.2.2 Pipeline Stress Computation

For buried pipelines, apart from the stress caused by an earthquake, there are other factors, such as the dead load, temperature change and Poisson's effect, to cause stress in tube body.

The dead load is produced by the soil covering the pipe. When the pipe is buried above the waterline, the dead load can be calculated as follows:

$$P_d = \sum \gamma_i H_i \quad (4.36)$$

where γ_i is the unit weight of the ith layer soil above the pipe, and H_i is and thickness of the ith layer soil above the pipe.

When the ground water needs to be considered, the dead load on pipelines is

$$P_d = \sum \gamma_i H_i - \gamma_w h \quad (4.37)$$

where γ_w is the unit weight of water and usually takes the value of 10kN/m^3, and h is the height of water above the pipe.

When the dead load is applied to a buried pipeline, according to the analysis model shown in Fig. 4.6, the hoop stress is

$$\sigma_{cs} = \frac{F_b}{A_w} + \frac{M_b t}{2I_w} \quad (4.38)$$

where A_w is the area of the pipe wall, I_w is the moment of inertia of the pipe wall, t is the thickness of the pipe wall, F_b and M_b are the axial force and bending moment of the pipe, respectively, and can be calculated as follows (Spangler, 1964):

$$F_b = K_f P_d r \quad (4.39)$$

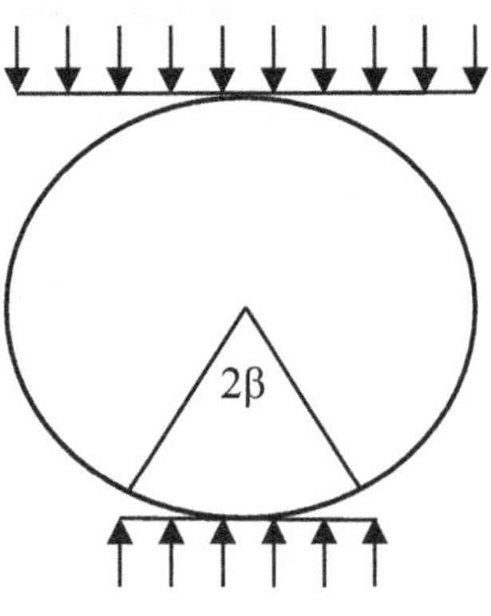

Fig. 4.6 The dead load on the pipeline

Tab 4.5 The coefficients for different β

$\beta(°)$	K_f	K_m
0	0.1061	0.5872
15	0.0990	0.4685
30	0.0796	0.3772
45	0.0531	0.3140
60	0.0265	0.2754
75	0.0071	0.2558
90	0.0000	0.2500

$$M_b = K_m P_d r \tag{4.40}$$

where r is the pipe radius, K_f and K_m are the coefficients and their values are shown in Table 4.5. Apparently, the axial force and bending moment decrease with the increase of β. Generally, $\beta = 15°$ is a good choice in engineering practice.

The internal pressure in the pipes, such as water pressure and gas pressure, can also cause hoop stress in the pipe. It is can be calculated by

$$\sigma_{cp} = \frac{p_I D}{2t} \tag{4.41}$$

where p_I is the internal pressure, D is pipe diameter.

Then the total hoop stress is

$$\sigma_c = \sigma_{cs} - \sigma_{cp} \tag{4.42}$$

When the temperature between the installation and the operation is different, the temperature stress may arise and can be calculated as follows:

$$\sigma_{lt} = -\alpha E(T_2 - T_1) \tag{4.43}$$

where α is the coefficient of the thermal expansion, E is the elastic modulus of pipe materials, T_1 is the installation temperature and T_2 is the operation temperature.

For a continuous pipeline, because the pipe is very long and the axial deformation is restrained, an axial stress is generated by Poisson's effect. It can be calculated by:

$$\sigma_{lp} = -\gamma \sigma_c \tag{4.44}$$

where γ is the Poisson's ratio.

The stress in pipeline is the combination of above stresses. For two-dimensional stresses, the von Mises yield criterion is usually used as the strength failure criterion. Therefore, the Mises stress of the pipeline can be written as follows:

$$\sigma_{eq} = \sqrt{\sigma_c^2 - \sigma_c\sigma_l + \sigma_l^2} \tag{4.45}$$

where

$$\sigma_l = \sigma_{lt} + \sigma_{lp} + \sigma_{le} + \sigma_{lb} \tag{4.46}$$

where σ_{le} is the axial stress caused by earthquake which can be calculated by (4.28), σ_{lb} is the bending stress caused by earthquake and can be given by

$$\sigma_{lb} = \frac{MD}{2I} \tag{4.47}$$

where M is the bending moment caused by the seismic wave and I is the moment of inertia of the pipe.

4.3 Seismic Response Analysis of Pipeline Networks

Besides the seismic analysis of a single pipeline, the seismic analysis of whole pipeline networks is necessary for precise analysis and design of buried pipeline system. This kind of analysis can be carried out by using the finite element method. As described in Sect. 4.2, a buried pipeline can usually be idealized as a beam on elastic foundation and its seismic responses can be obtained through the pseudo-static approach. For the model shown in Fig. 4.7, the axial and lateral motion equations can be described as follows:

$$EA\frac{\partial^2 u(x,t)}{\partial x^2} - k_A u(x,t) = -k_A u_g(x,t) \tag{4.48}$$

$$EI\frac{\partial^4 v(x,t)}{\partial x^2} - k_L v(x,t) = -k_L v_g(x,t) \tag{4.49}$$

where EA and EI are the axial and bending stiffness of the pipeline, respectively; k_A and k_L are the spring stiffness per unit length of the soil surrounding the pipeline along with the axial and lateral directions, respectively; $u(x, t)$ and $v(x, t)$ are the axial and lateral displacements of the pipeline, respectively; $u_g(x, t)$ and $v_g(x, t)$ are the axial and lateral displacements of ground motion, respectively.

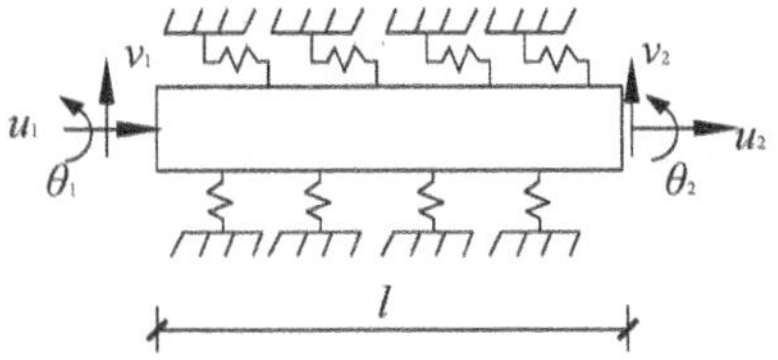

Fig. 4.7 Pipeline modeled as a beam on elastic foundation

Using the basic idea of finite element method (Zienkiewicz, Taylor, & Zhu, 2005; Liu, Sun, Miao, & Li, 2015a, Liu, Zhao, & Li, 2015b), the element stiffness matrix of a buried pipeline element can be described as follows:

$$[\mathbf{K_P}] = \begin{bmatrix} \frac{EA}{l} & 0 & 0 & -\frac{EA}{l} & 0 & 0 \\ & \frac{12EI}{l^3} & \frac{6EI}{l^2} & 0 & -\frac{12EI}{l^3} & \frac{6EI}{l^2} \\ & & \frac{4EI}{l} & 0 & -\frac{6EI}{l^2} & \frac{2EI}{l} \\ & \text{Symmetry} & & \frac{EA}{l} & 0 & 0 \\ & & & & \frac{12EI}{l^3} & \frac{6EI}{l^2} \\ & & & & & \frac{4EI}{l} \end{bmatrix} \tag{4.50}$$

where l is the length of element.

If the pipe-soil interaction is simulated as the axial and lateral springs, and the corresponding stiffness matrix can be written as follows (Wang, 1978):

$$[\mathbf{K_s}] = \begin{bmatrix} \frac{1}{3}\gamma & 0 & 0 & \frac{1}{6}\gamma & 0 & 0 \\ & \frac{13}{35}\lambda & \frac{11}{210}l\lambda & 0 & \frac{9}{70}\lambda & -\frac{13}{420}l\lambda \\ & & \frac{1}{105}l^2\lambda & 0 & \frac{13}{420}l\lambda & -\frac{1}{140}l^2\lambda \\ \text{Symmetry} & & & \frac{1}{3}\gamma & 0 & 0 \\ & & & & \frac{13}{35}\lambda & -\frac{11}{210}l\lambda \\ & & & & & \frac{1}{105}l^2\lambda \end{bmatrix} \tag{4.51}$$

where $\gamma = k_A l$; $\lambda = k_L l$.

For water supply networks, many pipe segments connect buried pipelines with joints. As shown in Fig. 4.8, a joint can be simulated by using a tension and compression spring in the axial direction, a bending spring in the rotating direction and a spring with infinite stiffness in the lateral direction. Then the element stiffness matrix of a joint can be described as follows:

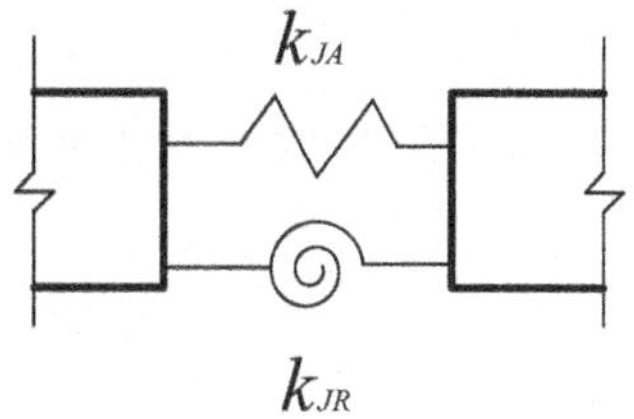

Fig. 4.8 Model of pipe joint between two pipeline segments

$$[\mathbf{K_J}] = \begin{bmatrix} k_{JA} & 0 & 0 & -k_{JA} & 0 & 0 \\ & k_{\infty} & 0 & 0 & -k_{\infty} & 0 \\ & & k_{JR} & 0 & 0 & -k_{JR} \\ & & & k_{JA} & 0 & 0 \\ & \text{Symmetry} & & & & \\ & & & & k_{\infty} & 0 \\ & & & & & k_{JR} \end{bmatrix} \tag{4.52}$$

where k_{JA} and k_{JR} represent the axial and bending springs stiffness of the joint, respectively; k_{∞} takes a large value and represents a lateral spring with infinite stiffness.

An axial joint spring obviously behaves differently when subjected to tension and compression and is usually described separately by a perfectly plastic model for tension and an elastic model for compression (O'Rourke and Liu, 1999). The relationship between the axial force and the displacement is shown in Fig. 4.9, where P_u and u_l are the ultimate axial resistant force and the ultimate axial deformation at the elastic phase in intention, respectively, and $u_{\max}$ is the maximal axial deformation in compression.

Considering that the element stiffness matrix $[\mathbf{K_P}]$, $[\mathbf{K_S}]$ and $[\mathbf{K_J}]$ are obtained in a local coordinate system, they must be transformed to the corresponding matrix in the global coordinate system before the element matrices are assembled to form the system stiffness matrix. The transformation matrix is expressed as follows:

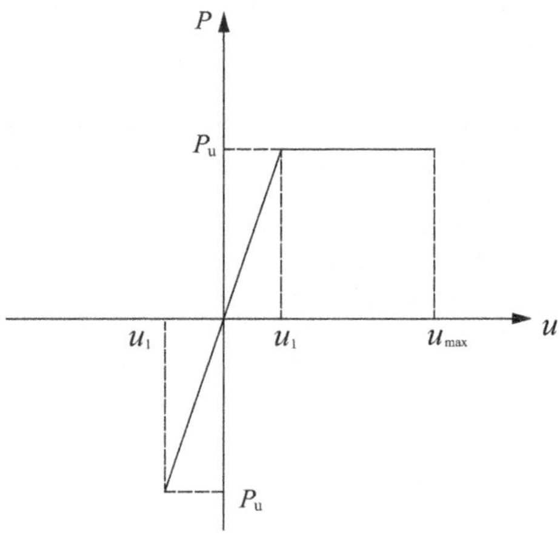

Fig. 4.9 The relationship between the axial force and axial deformation of a joint

$$[\mathbf{T}] = \begin{bmatrix} \cos\alpha & \sin\alpha & 0 & 0 & 0 & 0 \\ -\sin\alpha & \cos\alpha & 0 & 0 & 0 & 0 \\ 0 & 0 & 1 & 0 & 0 & 0 \\ 0 & 0 & 0 & \cos\alpha & \sin\alpha & 0 \\ 0 & 0 & 0 & -\sin\alpha & \cos\alpha & 0 \\ 0 & 0 & 0 & 0 & 0 & 1 \end{bmatrix} \tag{4.53}$$

where α represents the angle between local coordinate system and global coordinate system.

Then according to the principle of minimum potential energy, the basic finite element equation, which describes the relationship between the pipeline displacement and seismic wave excitations, can be established as follows:

$$[\mathbf{K}_{SYS}]\{\mathbf{u}(t)\} = [\mathbf{K}_S]\{\mathbf{x}_G(t)\} \tag{4.54}$$

where $[\mathbf{K}_{SYS}] = [\overline{\mathbf{K}}_P] + [\overline{\mathbf{K}}_S] + [\overline{\mathbf{K}}_J]$ is the system stiffness matrix in the global coordinate system, here the symbol "–" denotes the matrix in the global coordinate system; $\{\mathbf{u}(t)\}$ and $\{\mathbf{x}_G(t)\}$ are the displacement vectors of the pipeline and the ground motion in the global coordinate system, respectively.

Solve Eq (4.54), the displacements of the elements, deformations of the joints, strains and stresses of the pipes etc. can be obtained. For example, if element i is a joint, its axial deformation is expressed as follows:

$$u^i_{\text{joint}} = u^{i+1}_1 - u^{i-1}_2 \tag{4.55}$$

where u^{i+1}_1 is the displacement of the left node of element $i+1$; u^{i-1}_2 is the displacement of the right node of element i-1.

If element i is a pipeline element, the axial strain in the middle of the pipeline element could be expressed as follows:

$$\varepsilon^i_{pipe} = \frac{u^i_2 - u^i_1}{l} \tag{4.56}$$

where u^i_1 is the displacement of the left node; u^i_2 is the displacement of the right node.

It is valuable to point out that the above-modeled method has been verified by experiments, as shown in Fig. 4.10 (Liu, Miao, & Wang, 2017a; Liu, Xu, & Li, 2017b)

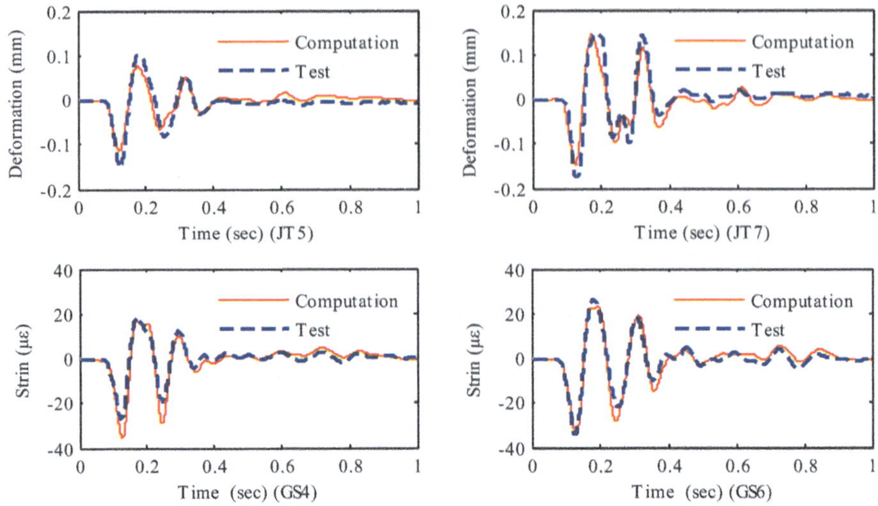

Fig. 4.10 Comparison between artificial earthquake experiments and FEA results

4.4 Seismic Reliability Evaluation of Buried Pipeline

Above analysis indicates that the pipeline deformation and inertial force under earthquake are mainly determined by earthquake intensity and physical parameter of site soil. However, these parameters are affected by many complex factors and often belong to random variables. Thus, a suitable method should be identified to analyze the seismic response of buried pipelines considering the randomness associated with these parameters. Obviously, the probability-based method is a suitable choice for this target.

4.4.1 Uncertainty of Pipeline Resistance

In fact, not only the seismic response of pipelines is random, the resistibility and deformability of pipeline joint are also random due to the uncertainties in the material properties and in the construction conditions. Table 4.6 presents the ultimate joint deformation of different pipelines given by Chinese researchers(Han, 2002). In the table, R_1, R_2 and P_k are the crack displacement, leakage displacement and the crack load of pipeline joints, respectively. A number of experiments indicate that the means and standard deviations of R_1 or R_2 vary little with diameters. Therefore, the results are available for the pipelines with different diameters (usually 100 mm–800 mm). Meanwhile, statistical analysis indicates R_1 and R_2 follow the normal distribution with 0.05 significant level.

Table 4.6 Pipeline joint crack, leakage displacement and crack load

Material	Joint material	R_1 (mm)		R_2 (mm)		P (kN)	
		Mean	Standard deviation	Mean	Standard deviation	Mean	Standard deviation
Cast iron	Asbestos cement	0.32	0.18	2.65	1.08	68.6	7.04
Cast iron	Self-stress cement	0.58	0.11	2.88	1.19	55.3	6.40
Cast iron	Gasket and asbestos ash	4.50	1.88	25.68	3.62	83.2	18.61
Cast iron	Gasket and self-stress ash	5.59	0.76	24.98	4.26	58.9	14.46
Reinforced concrete	Cement mortar	0.42	0.29	3.00	1.38	–	–
Prestressed concrete	Gasket	5.00	2.00	38.6	4.13	–	–

4.4.2 *Seismic Reliability Analysis of Buried Pipelines*

When the seismic response of a pipeline and its allowable resistance are known, the structural reliability method can be used to give pipeline reliability. Let the pipeline performance function be:

$$Z = g(R, S) = R - S \tag{4.57}$$

where g is a state function or performance function, R and S are the resistance and the loading effect of the pipeline, respectively.

For example, if the joint deformation of the pipeline is used to judge whether the pipeline is damaged or not, the performance function of pipeline can be written as follows:

$$Z_u = R_2 - S_u \tag{4.58}$$

where S_u is the joint deformation of the pipeline under earthquake, R_2 is the leakage displacement of the pipeline.

Apparently, when $Z_u < 0$, water will leak from joint, i.e. the pipeline is in the failure state. When $Z_u \geq 0$, however, no water leaks from joint, i.e. the pipeline is in the safe state. Usually, such kind of the two-state division is called the two-state damage criteria.

Assume R and S both follow the normal distributions. Then Z also follows normal distribution and its probability density function is denoted by:

$$f(z) = \frac{1}{\sqrt{2\pi}\sigma} e^{-\frac{(z-\mu)^2}{2\sigma^2}} \quad (4.59)$$

where μ and σ are the mean and the standard deviation of Z, respectively, and can be stated as follows:

$$\mu = \mu_R - \mu_S \quad (4.60)$$

$$\sigma = \sqrt{\sigma_R^2 + \sigma_S^2} \quad (4.61)$$

where μ_R and σ_R are the mean and standard deviation of R_2, μ_S and σ_S are the mean and standard deviation of S_u, respectively.

Then the failure probability of the pipeline is given by:

$$P_f = P(Z_u < 0) = \int_{-\infty}^{0} \frac{1}{\sqrt{2\pi}\sigma} e^{-\frac{(z-\mu)^2}{2\sigma^2}} \mathrm{d}z = \Phi\left(-\frac{\mu}{\sigma}\right) = \Phi(-\beta) \quad (4.62)$$

where $\Phi(\cdot)$ is standard normal distribution function and $\beta = \frac{\mu}{\sigma}$ is called as the reliability index.

Meanwhile, the reliability of the pipeline is:

$$P_s = 1 - P_f = \Phi\left(\frac{\mu}{\sigma}\right) \quad (4.63)$$

The reliability shown in Eq. (4.63) is the seismic reliability of one joint. In engineering practice, however, the seismic reliability of a pipeline with many joints should be considered. Then, a failure dependence problem of multi-joint pipelines should be solved. Many methods are proposed to deal with this problem. Without loss of generality, it is assumed a pipeline consists of n joints. Then the event E which means the pipeline failure is the union of the events E_i that each joint fails. That is:

$$E = \bigcup_{i=1}^{n} E_i \quad (4.64)$$

Let $P_f(\cdot)$ represent the probability that a failure event happens, it exists:

$$\max \quad P_f(E_i) \le P_f(E) \le \sum_{i=1}^{n} P_f(E_i) \quad (4.65)$$

The equality at the right side of Eq. (4.65) means the failure of all joints is *s*-independent, while that at the left side means the failure of all joints is completely dependent. Usually, the interval of Eq. (4.65) will become large quickly with the increase of n. In practice, Ditlevsen boundary estimation (Ditlevsen, 1979), which can give a narrow interval estimation, should be adopted. It is stated as follows:

$$P_{fL} \leq P_f(E) \leq P_{fu} \tag{4.66}$$

where

$$P_{fL} = P(E_1) + \max\left\{\sum_{i=2}^{n}\left[P(E_i) - \sum_{j=1}^{i-1} P_L(E_i E_j)\right], 0\right\} \tag{4.67}$$

$$P_{fu} = \sum_{i=1}^{n} P(E_i) - \sum_{i=1}^{n} \max_{j<i} P_R(E_i E_j) \tag{4.68}$$

$$P_L(E_i E_j) = q_i + q_j \tag{4.69}$$

$$P_R(E_i E_j) = \max(q_i, q_j) \tag{4.70}$$

$$q_i = \Phi(-\beta_i)\Phi\left(-\frac{\beta_j - \rho_{ij}\beta_i}{\sqrt{1-\rho_{ij}^2}}\right) \tag{4.71}$$

$$q_j = \Phi(-\beta_j)\Phi\left(-\frac{\beta_i - \rho_{ij}\beta_j}{\sqrt{1-\rho_{ij}^2}}\right) \tag{4.72}$$

where β_i is the reliability index of joint i, ρ_{ij} is the correlation coefficient of the damage between joints i and j.

Then, the pipeline failure probability can be calculated by the following point estimation.

$$P_f(E) = \frac{1}{2}(P_{fL} + P_{fu}) \tag{4.73}$$

It should be pointed out that the above analysis is on the premier that the pipeline damage is defined as the event that the pipeline joint deformation is larger than the allowance deformation, which is apparently suitable for segmented pipelines. For a continuous pipeline, such as the welded steel pipeline, however, the damage should be the event that the pipeline stress is larger than the material strength. The analytical method is similar to the above method and therefore will not to be described in detail.

References

ALA. (2001). Seismic fragility formulations for water systems. ASCE.

Ariman, T., & Muleski, G. E. (1981). A review of the response of buried pipelines under seismic excitations. *Earthquake Engineering Structural Dynamics, 9*(2), 133–152.

Ditlevsen, O. (1979). Narrow reliability bounds for structural systems. *Journal of Structural Mechanics, 7*(4), 453–472.
Han, Y. (2002). *Research on seismic reliability of urban underground pipe network system*. Ph. D. dissertation, Dalian university of technology, China (in Chinese).
Hou, Z. L. (1990). *Underground pipeline aseismatic*. Academic Books and Periodicals Press (in Chinese).
Hwang, H. H., Lin, H., Shinozuka, M. (1998). Seismic performance assessment of water delivery systems. *Journal of Infrastructure Systems, 4*(3), 118–125.
Kuesel, T. R. (1969). Earthquake design criteria for subways. *Journal of the Structural Division, ASCE ST6*, 1213–1231.
Li, J. (1992). *Earthquake disaster prediction and disaster prevention planning*. Henan Science and Technology Press (in Chinese).
Liu, W., Sun, Q. W., Miao, H. Q., & Li, J. (2015a). Nonlinear stochastic seismic analysis of buried pipeline systems. *Soil Dynamics and Earthquake Engineering, 2015*(74), 69–78.
Liu, W., Zhao, Y. G., & Li, J. (2015b). Seismic functional reliability analysis of water distribution networks. *Structure and Infrastructure Engineering, 11*(3), 363–375.
Liu, W., Miao, H. Q., Wang, C., & Li, J. (2017a). Experimental validation of a model for seismic simulation and interaction analysis of buried pipe networks. *Soil Dynamics and Earthquake Engineering, 100*, 113–130.
Liu, W., Xu, L., & Li, J. (2017b). Seismic design of water distribution networks using auto-generation techniques. *Structure and Infrastructure Engineering, 13*(7), 831–843.
Liu, H. X. (1986). Damage in the Great Earthquake in Tangshan. Beijing: *Seismological Press*. (in Chinese).
Newmark, N. M. (1967). Problems in wave propagation in soil and rock. In *Proceedings of International Symposium on Wave Propagation and Dynamics Properties of Earth Materials* (pp. 7–26), New Mexico.
O'Rourke, M. J., & Liu, X. (1999). *Response of buried pipelines subject to earthquake effects*. Buffalo: MCEER, University at Buffalo.
Sato, N., et al. (1988). Observation of seismic ground and buried pipe strain in a very dense seismometer array. In *The 9th World Conference on Earthquake Engineering*, Tokyo-Kyouto.
Shinozuka, M., & Koike, T. (1979). *Estimation of structural strains in underground lifeline pipes*. NewYork: Department of Civil Engineering and Engineering Mechanics.
Spangler, M. G. (1964). Pipeline crossings under railroads and highways. *Journal American Water Works Association, 56*(8), 1029–1046.
Sun, S. P. (1986). Seismic problem of buried pipeline. In *The Third Session of the National Aseismatic Working Conference* (in Chinese).
Wang, L. R. L. (1978). Vibration frequencies of buried pipelines. *Journal of the Technical Councils of ASCE, 104*(1), 71–89.
Wang, L. R. L., Sun. S., & Shen, S. (1985). *Seismic damage behavior of buried lifeline systems during recent severe earthquake in the U. S., China and other countries* (Technical Report No. ODU LEE-02 in Lifeline earthquake Engineering Research Series).
Zienkiewicz, O. C., Taylor, R. L., & Zhu, J. Z. (2005). *The finite element method: Its basis and fundamentals* (6th ed.). Butterworth-Heinemann.

Chapter 5
Seismic Response Analysis of Structures

In lifeline network systems, there are various kinds of structures, such as the high-voltage transmission towers in power systems, the pump house in water supply systems, etc. The seismic performance and aseismic reliability of these structures are of great importance for the safety assessment and resilience enhancement of lifeline network systems. This chapter will deal with the seismic response analysis of lifeline engineering structures.

5.1 Structural Analysis Model

5.1.1 General Finite Element Model

Finite element method (FEM) is a typical discretization method in which the structure is divided rationally into finite elements and its mechanical performance can be analyzed by such kind of model. The element divisions of truss, plate and solid structures are shown in Fig. 5.1.

The responses of a structure under external loads, such as stress, strain and displacement, could be expressed as a continuous function of the spatial coordinates. After the above discretization, the function is replaced approximately by a series of sectional continuous functions which are only smooth inside each element. The sectional continuous function is called the interpolation function. Figure 5.2 shows a rectangle element with four nodes *i*, *j*, *k* and *l*. For each node, two displacements are needed to delineate its deformation and therefore there are eight displacements in total. Correspondingly, a function with eight unknown coefficients can be adopted as the element's displacement function as follows:

$$\left.\begin{aligned} u &= a_1 + a_2 x + a_3 y + a_4 xy \\ v &= a_5 + a_6 x + a_7 y + a_8 xy \end{aligned}\right\} \tag{5.1}$$

J. Li and W. Liu, *Lifeline Engineering Systems*,
https://doi.org/10.1007/978-981-15-9101-3_5

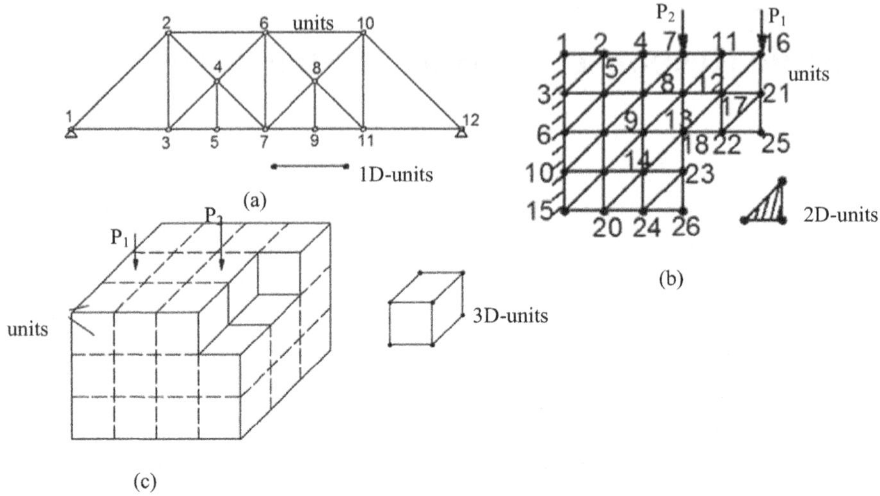

Fig. 5.1 Finite element divisions of structures

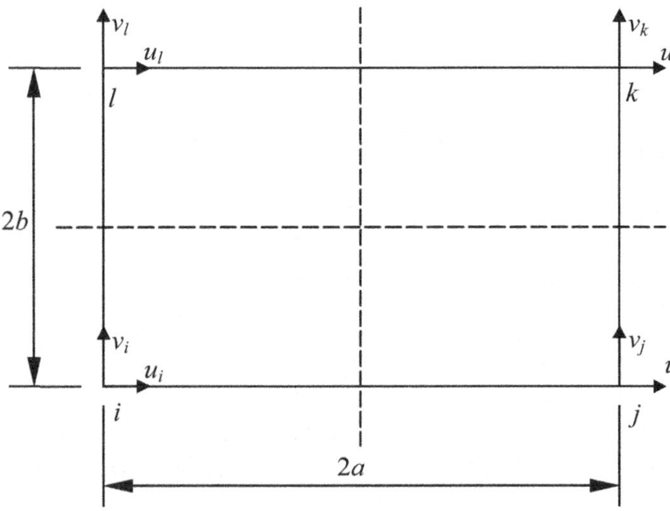

Fig. 5.2 A rectangle element

The eight unknown coefficients can be obtained based on the displacements of four nodes. In this way, Eq. (5.1) becomes

$$\begin{Bmatrix} u \\ v \end{Bmatrix} = \begin{bmatrix} N_i & 0 & N_j & 0 & N_k & 0 & N_l & 0 \\ 0 & N_i & 0 & N_j & 0 & N_k & 0 & N_l \end{bmatrix} \begin{Bmatrix} u_i \\ v_i \\ u_j \\ v_j \\ u_k \\ v_k \\ u_l \\ v_l \end{Bmatrix} \tag{5.2}$$

where N_i, N_j, N_k and N_l are called the shape functions and their expressions are

$$N_i = \frac{1}{4}\left(1 - \frac{x}{a}\right)\left(1 - \frac{y}{b}\right) \tag{5.3a}$$

$$N_j = \frac{1}{4}\left(1 + \frac{x}{a}\right)\left(1 - \frac{y}{b}\right) \tag{5.3b}$$

$$N_k = \frac{1}{4}\left(1 + \frac{x}{a}\right)\left(1 + \frac{y}{b}\right) \tag{5.3c}$$

$$N_l = \frac{1}{4}\left(1 - \frac{x}{a}\right)\left(1 + \frac{y}{b}\right) \tag{5.3d}$$

where a and b are the half-length and the half-width of the rectangle element and shown in Fig. 5.2.

In Eq. (5.2), the displacement of a point inside the element is expressed by displacements of nodes and the coordinate of the point.

In order to give the element energy, the strain-displacement relationship and the stress-strain relationship of the element should be introduced. For a point inside the element, the relationships between strains ε_x, ε_y, γ_{xy} and displacements u, v are

$$\varepsilon_x = \frac{\partial u}{\partial x} \tag{5.4a}$$

$$\varepsilon_y = \frac{\partial v}{\partial y} \tag{5.4b}$$

$$\gamma_{xy} = \frac{\partial u}{\partial y} + \frac{\partial v}{\partial x} \tag{5.4c}$$

Substituting Eq. (5.2) into (5.4) and introducing

$$\boldsymbol{\varepsilon} = \left(\varepsilon_x, \varepsilon_y, \gamma_{xy}\right)^{\mathrm{T}} \tag{5.5}$$

$$\mathbf{u} = (u_i, v_i, u_j, v_j, u_k, v_k, u_l, v_l)^{\mathrm{T}} \tag{5.6}$$

then Eq. (5.4a, 5.4b, 5.4c) can be rewritten as follows:

$$\boldsymbol{\varepsilon} = \mathbf{Bu} \tag{5.7}$$

where

$$\mathbf{B} = \begin{bmatrix} \frac{\partial N_i}{\partial x} & 0 & \frac{\partial N_j}{\partial x} & 0 & \frac{\partial N_k}{\partial x} & 0 & \frac{\partial N_l}{\partial x} & 0 \\ 0 & \frac{\partial N_i}{\partial y} & 0 & \frac{\partial N_j}{\partial y} & 0 & \frac{\partial N_k}{\partial y} & 0 & \frac{\partial N_l}{\partial y} \\ \frac{\partial N_i}{\partial y} & \frac{\partial N_i}{\partial x} & \frac{\partial N_j}{\partial y} & \frac{\partial N_j}{\partial x} & \frac{\partial N_k}{\partial y} & \frac{\partial N_k}{\partial x} & \frac{\partial N_l}{\partial y} & \frac{\partial N_l}{\partial x} \end{bmatrix} \tag{5.8}$$

is called as the strain matrix.

Meanwhile, the stress-strain relationship of a point inside the element is

$$\boldsymbol{\sigma} = \mathbf{D}\boldsymbol{\varepsilon} \tag{5.9}$$

where

$$\boldsymbol{\sigma} = \left(\sigma_x, \sigma_y, \tau_{xy}\right)^{\mathrm{T}} \tag{5.10}$$

D is an elastic matrix. For the plane stress problem,

$$\mathbf{D} = \frac{E}{1 - \nu^2}\begin{bmatrix} 1 & \nu & 0 \\ \nu & 1 & 0 \\ 0 & 0 & \frac{1-\nu}{2} \end{bmatrix} \tag{5.11}$$

where E and ν are the elastic modulus and Poisson ratio of medium, respectively.

Let V_e represent the strain energy of the element e. Then according to elasticity mechanics, it can be written as follows:

$$V_e = \frac{1}{2}\int_A \boldsymbol{\varepsilon}^{\mathrm{T}}\boldsymbol{\sigma} dA \tag{5.12}$$

where A is the integral region.

Substituting Eqs. (5.7) and (5.9) into the above equation, V_e can be written as follows:

$$V_e = \frac{1}{2}\int_A \mathrm{u}^{\mathrm{T}}\mathrm{B}^{\mathrm{T}}\mathrm{DBu} dA = \frac{1}{2}\mathrm{u}^{\mathrm{T}}\mathrm{K}_e\mathrm{u} \tag{5.13}$$

where

$$\mathrm{K}_e = \int_A \mathrm{B}^{\mathrm{T}}\mathrm{DB} dA \tag{5.14}$$

is the element stiffness matrix in the local coordinate.

During a dynamic process, each element of structures owns kinetic energy besides deformation energy. Assume that the velocity of a point inside the element is

$$\mathbf{v} = (\dot{u}, \dot{v})^{\mathrm{T}} \tag{5.15}$$

Considering Eq. (5.2), Eq. (5.15) becomes

$$\mathbf{v} = \mathbf{N}\mathbf{u} \tag{5.16}$$

where

$$\mathbf{N} = \left[\; N_i I_{2\times2} \; N_j I_{2\times2} \;\; N_k I_{2\times2} \; N_l I_{2\times2} \;\right] \tag{5.17}$$

where $I_{2\times2}$ is a 2×2 identity matrix

Then the kinetic energy of the element e, T_e, is

$$\begin{aligned} T_e &= \frac{1}{2}\int_A \mathrm{v}^{\mathrm{T}} \rho \mathrm{v} dA \\ &= \frac{1}{2}\int_A \dot{\mathrm{u}}^{\mathrm{T}} \mathrm{N}^{\mathrm{T}} \rho \mathrm{N} \dot{\mathrm{u}} dA \\ &= \frac{1}{2}\dot{\mathrm{u}}^{\mathrm{T}} \mathrm{M}_e \dot{\mathrm{u}} \end{aligned} \tag{5.18}$$

where ρ is the density and

$$\mathrm{M_e} = \int_A \mathrm{N}^{\mathrm{T}} \rho \mathrm{N} dA \tag{5.19}$$

is called the element mass matrix in the local coordinate.

Moreover, for a dynamic problem, the damping cannot be neglected. Assuming the damping force of a point is proportional to its velocity with a proportional coefficient η, the dissipation function of the element e can be written as follows:

$$\begin{aligned} R_e &= \frac{1}{2}\int_A \mathbf{v}^{\mathrm{T}} \eta \mathbf{v} dA \\ &= \frac{1}{2}\int_A \dot{\mathbf{u}}^{\mathrm{T}} \mathbf{N}^{\mathrm{T}} \eta \mathbf{N} \dot{\mathbf{u}} dA \\ &= \frac{1}{2}\dot{\mathbf{u}}^{\mathrm{T}} \mathbf{C}_e \dot{\mathbf{u}} \end{aligned} \tag{5.20}$$

where

$$\mathbf{C_e} = \int_A \mathbf{N}^{\mathrm{T}} \eta \mathbf{N} dA \tag{5.21}$$

is called the element damping matrix in the local coordinate.

Let the node force of an element be:

$$\mathbf{P} = (p_{ix}, p_{iy}, p_{jx}, p_{jy}, p_{kx}, p_{ky}, p_{lx}, p_{ly})^{\mathrm{T}} \tag{5.22}$$

Then the load potential energy of the element e is:

$$W_e = -\mathbf{u}^{\mathrm{T}}\mathbf{P} \tag{5.23}$$

So the energy function for the discrete element e can be given as follows:

$$\prod\nolimits_e = \int_0^t [T_e - (V_e + W_e) - R_e]dt \tag{5.24}$$

According to the Hamilton principle, the actual element displacement should give the extreme value of the above energy. Therefore, there exists

$$\delta \prod\nolimits_e = 0 \tag{5.25}$$

Using the variational operation and considering the arbitrariness of $\delta \boldsymbol{u}$, it can derive

$$\mathbf{M}_e\ddot{\mathbf{U}} + \mathbf{C}_e\dot{\mathbf{U}} + \mathbf{K}_e\mathbf{U} = \mathbf{P} \tag{5.26}$$

Repeating the above process for all elements and assembling in the global coordinate, a dynamic equation based on the FEM can be deduced as follows:

$$\mathbf{M}\ddot{\mathbf{U}} + \mathbf{C}\dot{\mathbf{U}} + \mathbf{K}\mathbf{U} = \mathbf{F} \tag{5.27}$$

where

$$\mathbf{M} = \sum_e \mathbf{T}_e^{\mathrm{T}}\mathbf{M}_e\mathbf{T}_e \tag{5.28a}$$

$$\mathbf{C} = \sum_e \mathbf{T}_e^{\mathrm{T}}\mathbf{C}_e\mathbf{T}_e \tag{5.28b}$$

$$\mathbf{K} = \sum_e \mathbf{T}_e^{\mathrm{T}}\mathbf{K}_e\mathbf{T}_e \tag{5.28c}$$

$$\mathbf{F} = \mathbf{T}_e^{\mathrm{T}}\mathbf{P} \tag{5.28d}$$

are the mass, damping, stiffness matrixes and load vector of the structure, respectively. $\mathbf{T}_{\mathrm{e}}$ is the element location matrix. $\ddot{\mathbf{U}}$, $\dot{\mathbf{U}}$ and $\mathbf{U}$ are the absolute acceleration, velocity and displacement, respectively.

Table 5.1 Elastic matrix of element

Element type		Elastic matrix
One-dimension	Bar	EA
	Pure bending beam	EI
Two-dimension	Plane stress	$\frac{E}{1-v^2}\begin{bmatrix} 1 & v & 0 \\ v & 1 & 0 \\ 0 & 0 & \frac{1-v}{2} \end{bmatrix}$
	Plane strain	$\frac{E}{(1-v)(1-2v)}\begin{bmatrix} 1-v & v & 0 \\ v & 1-v & 0 \\ 0 & 0 & \frac{1-2v}{2} \end{bmatrix}$
	Slab	$\frac{Et^3}{12(1-v^2)} = \begin{bmatrix} 1 & v & 0 \\ v & 1 & 0 \\ 0 & 0 & \frac{1-v}{2} \end{bmatrix}$
Three-dimension	$\frac{E}{(1-v)(1-2v)}\begin{bmatrix} 1-v & v & v & 0 & 0 & 0 \\ v & 1-v & v & 0 & 0 & 0 \\ v & v & 1-v & 0 & 0 & 0 \\ 0 & 0 & 0 & \frac{1-2v}{2} & 0 & 0 \\ 0 & 0 & 0 & 0 & \frac{1-2v}{2} & 0 \\ 0 & 0 & 0 & 0 & 0 & \frac{1-2v}{2} \end{bmatrix}$	

For general structural systems, the FEM equations are similar. The difference lies in the interpolation function and the elastic matrix. For different elements, the elastic matrix is shown in Table 5.1.

When the seismic ground motion is considered, the analytical equation of a linear structure can be expressed as follows:

$$\mathbf{M}\ddot{\mathbf{U}} + \mathbf{C}\dot{\mathbf{U}} + \mathbf{K}\mathbf{U} = -\mathbf{M}\ddot{\mathbf{U}}_g \tag{5.29}$$

where $\ddot{\mathbf{U}}_g$ is the ground motion in three directions.

5.1.2 Seismic Analysis Model of Structure-Equipment Systems

Structure-equipment systems could be regarded as one special category of lifeline engineering structures. Although the general FEM can be used to analyze such systems, sub-structure coupling method seems more suitable for considering the

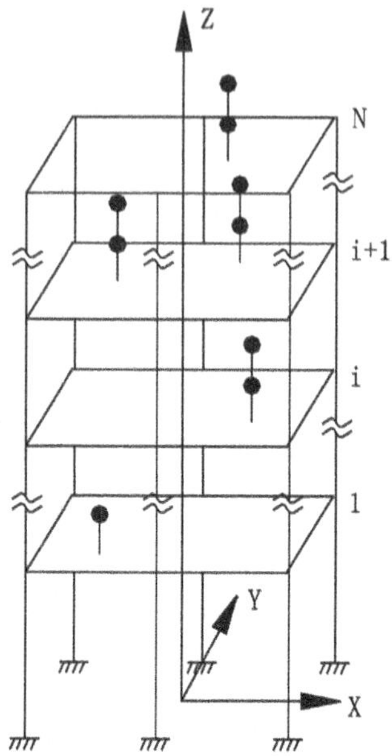

Fig. 5.3 Structure-equipment system model

specialty of the problem, and especially to deal with the different damping among structures and equipments.

Without loss of generality, a structure-equipment system shown in Fig. 5.3 is considered and the system is assumed to subject to the ground motion in two horizontal directions. Denote the main structures as P and assume that all slabs are of rigidity. The equipment on each floor is regarded as a sub-structure, S_i. Then, system kinetic energy can be expressed as follows:

$$\mathbf{T} = \frac{1}{2}\dot{\mathbf{x}}_P^{\mathrm{T}}\mathbf{M}_P\dot{\mathbf{x}}_P + \sum_{i=1}^{n}\frac{1}{2}\dot{\mathbf{x}}_{S_i}^{\mathrm{T}}\mathbf{M}_{S_i}\dot{\mathbf{x}}_{S_i} \tag{5.30}$$

where $\mathbf{M}_P$ is the mass matrix of the main structure, $\mathbf{M}_{S_i}$ is the mass matrix of the sub-structure S_i, These mass matrices can be lumped mass matrices or consistent mass matrices, n is the number of sub-structures, $\dot{\mathbf{x}}_P$ and $\dot{\mathbf{x}}_{S_i}$ are the absolute velocity vectors of the main structure and the sub-structure S_i and can be expressed as follows:

$$\dot{\mathbf{x}}_P = \dot{\mathbf{x}}_{gP} + \dot{\mathbf{U}}_P \tag{5.31a}$$

$$\dot{\mathbf{x}}_{S_i} = \dot{\mathbf{x}}_{gS_i} + \dot{\mathbf{U}}_{S_i} \tag{5.31b}$$

where $\dot{\mathbf{U}}_P$ and $\dot{\mathbf{U}}_{S_i}$ are the nodal relative velocity vector of the main structure and the sub-structure $S_{i,}$ respectively, $\dot{\mathbf{x}}_{gP} = \left(\dot{x}_g, \ldots, \dot{x}_g; \dot{y}_g, \ldots, \dot{y}_g; 0, \ldots, 0\right)^{\mathrm{T}}$

is the ground velocity vector and 0 means there is no torsion input, $\dot{\mathbf{x}}_{gS_i} = (\dot{x}_g, \ldots, \dot{x}_g; \dot{y}_g, \ldots, \dot{y}_g)^{\mathrm{T}}$.

For linear elastic systems, the potential energy can be expressed as follows:

$$V = V_P + \sum_{i=1}^{n} V_{S_i} \tag{5.32}$$

where

$$V_P = \frac{1}{2}\mathbf{U}_P^{\mathrm{T}}\mathbf{K}_P\mathbf{U}_P \tag{5.33}$$

$$V_{S_i} = \frac{1}{2}\bar{\mathbf{U}}_{S_i}^{\mathrm{T}}\bar{\mathbf{K}}_{S_i}\bar{\mathbf{U}}_{S_i} \tag{5.34}$$

where $\mathbf{K}_P$ is the stiffness matrix of the main structure and can be given by using the general FEM. After the boundary conditions are introduced, $\mathbf{K}_P$ becomes a nonsingular matrix. $\bar{\mathbf{K}}_{S_i}$ is the stiffness matrix of the sub-structure, which would be a singular matrix since the boundary conditions are not introduced. In fact, the structure corresponding to this matrix is a suspension sub-structure.

$$\mathbf{U}_P = \left(\mathbf{u}_1^{\mathrm{T}}, \ldots, \mathbf{u}_N^{\mathrm{T}}\right)^{\mathrm{T}} \tag{5.35a}$$

$$\tilde{\mathbf{U}}_{S_i} = \left(\mathbf{u}_{i0}^{\mathrm{T}}, \mathbf{u}_{i1}^{\mathrm{T}}, \ldots, \mathbf{u}_{iN_i}^{\mathrm{T}}\right)^{\mathrm{T}} \tag{5.35b}$$

where $\mathbf{u}_1^{\mathrm{T}} = (u_i, v_i, \theta_i)$ $(i = 1, 2, \ldots, N)$, $\mathbf{u}_{ij}^{\mathrm{T}} = \left(u_{ij}, v_{ij}\right)$ $(i = 1, 2, \ldots, N; j = 0, 1, 2, \ldots, N_i)$, $\mathbf{U}_P$ and $\tilde{\mathbf{U}}_{S_i}$ are the displacements related to the ground of the main structure and sub-structure, respectively, N is the number of structure floors, N_i is the number of lumped mass of sub-structures on the ith floor, u_{i0} and v_{i0} are the corresponding displacements of the main structure at the place of the origin of the sub-structure.

In order to remove the singularity of the stiffness matrix in Eq. (5.35), the connection conditions between main structure and sub-structures are introduced:

$$u_{i0} = u_{iP} - y_{S_i}\theta_i \tag{5.36a}$$

$$v_{i0} = v_{iP} + x_{S_i}\theta_i \tag{5.36b}$$

where u_{iP} and v_{iP} are the two-direction displacements at the mass center of the ith floor of main structure, θ_i is the rotation angle of the ith floor of main structure, x_{S_i} and y_{S_i} are the coordinates of the mass center of the ith floor.

Then Eq. (5.35b) can be rewritten as follows:

$$\bar{\mathbf{U}}_{S_i} = \begin{pmatrix} \mathbf{P}_{bi}\mathbf{T}_i\mathbf{U}_P \\ \mathbf{U}_{Si} \end{pmatrix} \tag{5.37}$$

where

$$\mathbf{P}_{bi} = \begin{bmatrix} 1 \; 0 \; -y_{S_i} \\ 0 \; 1 \quad x_{S_i} \end{bmatrix} \tag{5.38}$$

is called the coupling matrix of the main structure and sub-structure. And

$$\mathbf{T}_{\mathrm{i}} = \begin{bmatrix} \mathbf{T}_{\mathrm{i1}} \\ \mathbf{T}_{\mathrm{i2}} \\ \mathbf{T}_{\mathrm{i3}} \end{bmatrix} \tag{5.39}$$

is the layer positional matrix,

$$\mathbf{T}_{\mathrm{ij}} = \underset{3i+j}{[0, \ldots, 0, 1, 0, \ldots, 0]_{1\times 3N}} \tag{5.39a}$$

$$\mathbf{U}_{S_i} = \left(\mathbf{u}_{i1}^{\mathrm{T}}, \mathbf{u}_{i2}^{\mathrm{T}}, \ldots, \mathbf{u}_{iN_i}^{\mathrm{T}}\right)^{\mathrm{T}} \tag{5.40}$$

where $\mathbf{u}_{\mathrm{ij}}^{\mathrm{T}} = \left(u_{ij}, v_{ij}\right)$ $(i = 1, 2, \ldots, N; j = 0, 1, 2, \ldots, N_i)$.

Moreover, the stiffness matrix of the sub-structure $\tilde{\mathbf{K}}_{S_i}$ can be decomposed into blocks.

$$\tilde{\mathbf{K}}_{S_i} = \begin{bmatrix} \mathbf{K}_{bi} & \mathbf{K}_{bS_i} \\ \mathbf{K}_{S_ib} & \mathbf{K}_{S_i} \end{bmatrix} \tag{5.41}$$

where $\mathbf{K}_{S_i}$ is the stiffness matrix of the sub-structure fixed in the main structure and can be given using the general FEM. For example, the stiffness matrix of the storey-shear structure can be written as follows:

$$\mathbf{K}_{bi} = \begin{bmatrix} \mathbf{k}_{x_i1} & 0 \\ 0 & \mathbf{k}_{y_i1} \end{bmatrix} \tag{5.41a}$$

$$\mathbf{K}_{bS_i} = \begin{bmatrix} -K_{x_i1}, 0, \ldots, 0 & 0_{1\times N_i} \\ 0_{1\times N_i} & -K_{y_i1}, 0, \ldots, 0 \end{bmatrix} \tag{5.41b}$$

$$\mathbf{K}_{S_ib} = \mathbf{K}_{bS_i} \tag{5.41c}$$

Substituting Eqs. (5.37) and (5.41) into Eq. (5.34), the potential energy of substructures can be derived

$$V_{S_i} = \frac{1}{2}(\mathbf{U}_P^{\mathrm{T}}\mathbf{T}_i^{\mathrm{T}}\mathbf{P}_{bi}^{\mathrm{T}}\mathbf{K}_{bi}\mathbf{P}_{bi}\mathbf{T}_i\mathbf{U}_P + \mathbf{U}_P^{\mathrm{T}}\mathbf{T}_i^{\mathrm{T}}\mathbf{P}_{bi}^{\mathrm{T}}\mathbf{K}_{bS_i}\mathbf{U}_{S_i}$$

$$+ \mathbf{U}_{S_i}^{\mathbf{T}} \mathbf{K}_{S_i i} \mathbf{P}_{bi} \mathbf{T}_i \mathbf{U}_P + \mathbf{U}_{S_i}^{\mathbf{T}} \mathbf{K}_{S_i} \mathbf{U}_{S_i}\big) \tag{5.42}$$

The main difference between structure-equipment systems and general structures is the damping of the main structure and equipments. General FEM cannot simulate this difference while the sub-structure method can be used to solve the problem conveniently. In fact, when the proportional damping model is adopted, the system damping energy can be stated approximately as follows:

$$R = \frac{1}{2} \dot{\mathbf{U}}_{S_i}^{\mathbf{T}} \mathbf{K}_{S_i} \dot{\mathbf{U}}_{S_i} + \frac{1}{2} \sum_{i=1}^{n} \dot{\mathbf{U}}_{S_i}^{\mathbf{T}} \mathbf{K}_{S_i} \dot{\mathbf{U}}_{S_i} \tag{5.43}$$

Substitute kinetic energy, potential energy and damping energy into the Lagrange equation

$$\frac{d}{dt}\left(\frac{\partial T}{\partial \dot{\mathbf{U}}}\right) + \frac{\partial T}{\partial \mathbf{U}} = -\frac{\partial R}{\partial \dot{\mathbf{U}}} \tag{5.44}$$

Then the dynamic equation of the structure-equipment system can be given as follows (Li et al. 1997):

$$\mathbf{M}_P \ddot{\mathbf{U}}_P + \mathbf{C}_P \dot{\mathbf{U}}_P + \big(\mathbf{K}_P + \bar{\mathbf{K}}_P\big) \mathbf{U}_P + \sum_{i=1}^{n} \mathbf{K}_{PS_i} \mathbf{U}_{S_i} = -\mathbf{M}_P \ddot{\mathbf{U}}_{gP} \tag{5.45a}$$

$$\mathbf{M}_{S_i} \ddot{\mathbf{U}}_{S_i} + \mathbf{C}_{S_i} \dot{\mathbf{U}}_{S_i} + \mathbf{K}_{S_i P} \mathbf{U}_P + \mathbf{K}_{S_i} \mathbf{U}_{S_i} = -\mathbf{M}_{S_i} \ddot{\mathbf{U}}_{gS_i} \quad (i = 1, 2, \cdots, N) \tag{5.45b}$$

where

$$\bar{\mathbf{K}}_P = \mathbf{T}_i^{\mathbf{T}} \mathbf{P}_{bi}^{\mathbf{T}} \mathbf{K}_{bi} \mathbf{P}_{bi} \mathbf{T}_i \tag{5.46}$$

$$\mathbf{K}_{PS_i} = \mathbf{T}_i^{\mathbf{T}} \mathbf{P}_{bi}^{\mathbf{T}} \mathbf{K}_{bS_i} \tag{5.47}$$

$$\mathbf{K}_{S_i P} = \mathbf{K}_{bS_i} \mathbf{P}_{bi} \mathbf{T}_i = \mathbf{K}_{PS_i} \tag{5.48}$$

5.1.3 Dynamic Analysis Model of Structure Subject to Multi-point Ground Motions

For the structures which cover a long distance, such as long-span bridges, tunnels and buried pipelines, the seismic ground motion is different along the length of structures. As a result, the rigid basement assumption used in general structures is not suitable for such type of structures. Therefore, a special dynamic analysis model considering multi-point ground motions should be established.

As described in Sect. 5.1.1, the dynamic equation of a structure with multi-degree of freedom on which the external load $\mathbf{P}(t)$ acts can be written as follows:

$$\mathbf{M}\ddot{\mathbf{U}} + \mathbf{C}\dot{\mathbf{U}} + \mathbf{K}\mathbf{U} = \mathbf{P}(t) \tag{5.49}$$

When earthquakes occur, structural bearings move with the ground. If the subscript a represents the part related to the non-bearing nodes and subscript b represents the part related to the bearing nodes, Eq. (5.49) can be rewritten as follows:

$$\begin{bmatrix} \mathbf{M}_{aa} & \mathbf{M}_{ab} \\ \mathbf{M}_{ba} & \mathbf{M}_{bb} \end{bmatrix} \begin{Bmatrix} \ddot{\mathbf{U}}_a \\ \ddot{\mathbf{U}}_b \end{Bmatrix} + \begin{bmatrix} \mathbf{C}_{aa} & \mathbf{C}_{ab} \\ \mathbf{C}_{ba} & \mathbf{C}_{bb} \end{bmatrix} \begin{Bmatrix} \dot{\mathbf{U}}_a \\ \dot{\mathbf{U}}_b \end{Bmatrix} + \begin{bmatrix} \mathbf{K}_{aa} & \mathbf{K}_{ab} \\ \mathbf{K}_{ba} & \mathbf{K}_{bb} \end{bmatrix} \begin{Bmatrix} \mathbf{U}_a \\ \mathbf{U}_b \end{Bmatrix} = \begin{Bmatrix} \mathbf{0} \\ \mathbf{P}_b \end{Bmatrix} \tag{5.50}$$

According to the first row in the above equation, there is:

$$\mathbf{M}_{aa}\ddot{\mathbf{U}}_a + \mathbf{C}_{aa}\dot{\mathbf{U}}_a + \mathbf{K}_{aa}\mathbf{U}_a = -\left(\mathbf{M}_{ab}\ddot{\mathbf{U}}_b + \mathbf{C}_{ab}\dot{\mathbf{U}}_b + \mathbf{K}_{ab}\mathbf{U}_b\right) \tag{5.51}$$

Apparently, the displacement at any bearing nodes will cause the displacements of all non-restriction nodes. Let the number of structural bearing nodes be n and each bearing node owns three translation freedoms. Then the total degree of freedom is $L = 3n$. Let the displacement of the freedom l is $u_{sb,l}$ and the pseudo-static displacements of all other non-bearing nodes caused by $u_{sb,l} = 1$ are the vector $\mathbf{r}_{ab,l}$. Then the pseudo-static displacements of all other non-bearing nodes caused by the move of bearing nodes are denoted by:

$$\mathbf{U}_{sa} = \sum_{l=1}^{L} \mathbf{r}_{ab,l} u_{sb,l} \tag{5.52}$$

Apparently, the total displacement of structural node is the summation of pseudo-static displacement and dynamic relative displacement. As bearing nodes move with the ground, their dynamic relative displacements are zero. That is

$$\mathbf{U}_a = \mathbf{U}_{sa} + \mathbf{U}_{da} \tag{5.53a}$$

$$\mathbf{U}_b = \mathbf{U}_{sb} \tag{5.53b}$$

Substituting the above two equations into Eq. (5.51), it will give

$$\begin{aligned} \mathbf{M}_{aa}\ddot{\mathbf{U}}_{da} + \mathbf{C}_{aa}\dot{\mathbf{U}}_{da} + \mathbf{K}_{aa}\mathbf{U}_{da} &= -\left(\mathbf{M}_{aa}\ddot{\mathbf{U}}_{sa} + \mathbf{M}_{ab}\ddot{\mathbf{U}}_{sb}\right) \\ &- \left(\mathbf{C}_{aa}\dot{\mathbf{U}}_{sa} + \mathbf{C}_{ab}\dot{\mathbf{U}}_{sb}\right) - \left(\mathbf{K}_{aa}\mathbf{U}_{sa} + \mathbf{K}_{ab}\mathbf{U}_{sb}\right) \end{aligned} \tag{5.54}$$

It can be proved that the third term of the right part of equation is equal to zero at any instant of time. Meanwhile, the second term of the right part of equation can be neglected because it is usually very small. Thus, Eq. (5.51) can be rewritten as follows:

$$\mathbf{M}_{aa}\ddot{\mathbf{U}}_{da} + \mathbf{C}_{aa}\dot{\mathbf{U}}_{da} + \mathbf{K}_{aa}\mathbf{U}_{da} = -\left(\mathbf{M}_{aa}\ddot{\mathbf{U}}_{sa} + \mathbf{M}_{ab}\ddot{\mathbf{U}}_{sb}\right) \tag{5.55}$$

Furthermore, it is noticed that the following equation is correct at any time, there exists

$$\mathbf{K}_{aa}\mathbf{U}_{sa} + \mathbf{K}_{ab}\mathbf{U}_{sb} = 0 \tag{5.56}$$

Thus, the relationship between $\ddot{\mathbf{U}}_{sa}$ and $\ddot{\mathbf{U}}_{sb}$ can be established as:

$$\ddot{\mathbf{U}}_{sa} = -\mathbf{K}_{aa}^{-1}\mathbf{K}_{ab}\ddot{\mathbf{U}}_{sb} \tag{5.57}$$

Substituting Eq. (5.57) into Eq. (5.55) will give:

$$\mathbf{M}_{aa}\ddot{\mathbf{U}}_{da} + \mathbf{C}_{aa}\dot{\mathbf{U}}_{da} + \mathbf{K}_{aa}\mathbf{U}_{da} = \left(\mathbf{M}_{aa}\mathbf{K}_{aa}^{-1}\mathbf{K}_{ab} - \mathbf{M}_{ab}\right)\ddot{\mathbf{U}}_{sb} \tag{5.58}$$

Above equation is a dynamic equation with input acceleration at bearing nodes, which can be seen as the external load.

Noticing that for the lumped mass system, $\mathbf{M}_{ab} = \mathbf{0}$. $\mathbf{M}_{aa}$ and $\mathbf{K}_{aa}$ are the mass and stiffness matrices without considering the bearing relative displacement. Using Reyleigh damping assumption, $\mathbf{C}_{aa}$ can be expressed as the linear combination of $\mathbf{M}_{aa}$ and $\mathbf{K}_{aa}$, $\mathbf{U}_{da}$ represents the dynamic relative displacement of non-bearing nodes. Therefore all the subscript in Eq. (5.58) can be removed. Moreover, let $\mathbf{K}_g$ represent elastic coupling matrix $\mathbf{K}_{ab}$ and $\ddot{\mathbf{U}}_g$ represent the acceleration at bearing node, i.e.

$$\ddot{\mathbf{U}}_g = \left(\ddot{u}_{g1}, \ddot{u}_{g1}, \ldots, \ddot{u}_{gL}\right) \tag{5.59}$$

Then, for a lumped mass system, Eq. (5.58) can be expressed as follows:

$$\mathbf{M}\ddot{\mathbf{U}} + \mathbf{C}\dot{\mathbf{U}} + \mathbf{K}\mathbf{U} = \mathbf{M}\mathbf{K}^{-1}\mathbf{K}_g\ddot{\mathbf{U}}_g \tag{5.60}$$

Above equation is the structural dynamic analysis equation under multi-point ground motion.

5.2 Deterministic Seismic Response Analysis of Structures

Deterministic analysis means that the ground motion is regarded as a deterministic time process and the structural parameters, such as mass, stiffness and damping, are also deterministic variables. Generally, for such kind of analysis, the basic analysis methods can be classified into time history analysis, frequency domain analysis and vibration modes superposition methods. Considering the frequency domain analysis method and modes superposition method are limited in the analysis of nonlinear structural systems, only time history analysis method is introduced herein.

5.2.1 Linear Acceleration Algorithm

Taking Eq. (5.29) as example, at the time of $t_{j+1} = t_j + \Delta t$, structure dynamic equation is

$$\mathbf{M}\ddot{\mathbf{U}}_{j+1} + \mathbf{C}\dot{\mathbf{U}}_{j+1} + \mathbf{K}\mathbf{U}_{j+1} = -\mathbf{M}\ddot{\mathbf{U}}_{g,j+1} \tag{5.61}$$

Linear acceleration algorithm assumes the acceleration between t_j and t_{j+1} is a linear function of time interval τ and its slope can be given by the value of two ends. That is

$$\dddot{\mathbf{U}}_j = \frac{\ddot{\mathbf{U}}_{j+1} - \ddot{\mathbf{U}}_j}{\Delta t} = \frac{\Delta\ddot{\mathbf{U}}_j}{\Delta t} \tag{5.62}$$

Meanwhile, in the time interval (t_j, t_{j+1}), the velocity and displacement of any time can be given by the following expressions:

$$\dot{\mathbf{U}}(t_j + \tau) = \dot{\mathbf{U}}_j + \ddot{\mathbf{U}}_j\tau + \frac{1}{2}\dddot{\mathbf{U}}_j\tau^2 \tag{5.63}$$

$$\mathbf{U}(t_j + \tau) = \mathbf{U}_j + \dot{\mathbf{U}}_j\tau + \frac{1}{2}\ddot{\mathbf{U}}_j\tau^2 + \frac{1}{6}\dddot{\mathbf{U}}_j\tau^3 \tag{5.64}$$

When $\tau = \Delta t$, substituting Eq. (5.62) into Eqs. (5.63) and (5.64), the velocity and displacement at the time of t_{j+1} are:

$$\ddot{\mathbf{U}}_{j+1} = \dot{\mathbf{U}}_j + \frac{1}{2}\ddot{\mathbf{U}}_j\Delta t + \frac{1}{2}\ddot{\mathbf{U}}_{j+1}\Delta t \tag{5.65}$$

$$\mathbf{U}_{j+1} = \mathbf{U}_j + \dot{\mathbf{U}}_j\Delta t + \frac{1}{3}\ddot{\mathbf{U}}_j\Delta t^2 + \frac{1}{6}\ddot{\mathbf{U}}_{j+1}\Delta t^2 \tag{5.66}$$

Substituting above equations into Eq. (5.61) will give:

$$\bar{\mathbf{K}}\ddot{\mathbf{U}}_{j+1} = \bar{\mathbf{P}}_{j+1} \tag{5.67}$$

where

$$\tilde{\mathbf{K}} = \mathbf{M} + \frac{\Delta t}{2}\mathbf{C} + \frac{\Delta t^2}{6}\mathbf{K} \tag{5.68}$$

is called pseudo-static stiffness matrix, and

$$\bar{\mathbf{P}}_{j+1} = -\mathbf{M}\ddot{\mathbf{U}}_{g,j+1} - \mathbf{C}\left(\dot{\mathbf{U}}_j + \frac{1}{2}\ddot{\mathbf{U}}_j\Delta t\right) - \mathbf{K}\left(\mathbf{U}_j + \dot{\mathbf{U}}_j\Delta t + \frac{1}{3}\ddot{\mathbf{U}}_j\Delta t^2\right) \quad (5.69)$$

is called the pseudo-static load vector.

Generally, Eq. (5.67) is called the pseudo-static equation. Combining the initial conditions and time history of seismic ground motions, the pseudo-static equation can be solved easily. Meanwhile, corresponding velocity and displacement can be given using Eqs. (5.65) and (5.66).

The method presented above is suitable for dynamic analysis of linear elastic structure. When structural performance is in non-linear state, the incremental analysis is needed. For such case, the dynamic equation at the time t_j and t_{j+1} is given by:

$$\mathbf{M}\ddot{\mathbf{U}}_j + \mathbf{C}\dot{\mathbf{U}}_j + \mathbf{F}_j = -\mathbf{M}\ddot{\mathbf{U}}_{g,j} \quad (5.70)$$

$$\mathbf{M}\ddot{\mathbf{U}}_{j+1} + \mathbf{C}\dot{\mathbf{U}}_{j+1} + \mathbf{F}_{j+1} = -\mathbf{M}\ddot{\mathbf{U}}_{g,j+1} \quad (5.71)$$

Comparing with the dynamic equation of linear elastic structures, here the restoring force vector **F** replaces the elastic force vector **KU** in Eqs. (5.61). Introducing incremental stiffness method in non-linear analysis, the non-linear incremental dynamic equation can be established. To this end, let us consider a single-degree-of-freedom structure firstly.

The relationship between resorting force and displacement is shown in Fig. 5.4. Then in a suitable interval, the following approximate expression can be established

$$f_{j+1} = f_j + \bar{K}_j \Delta U_j \quad (5.72)$$

where $\bar{K}_j$ is the tangent stiffness at j.

When time incremental Δt is suitable, above incremental accumulation can give non-linear response with relatively high accuracy.

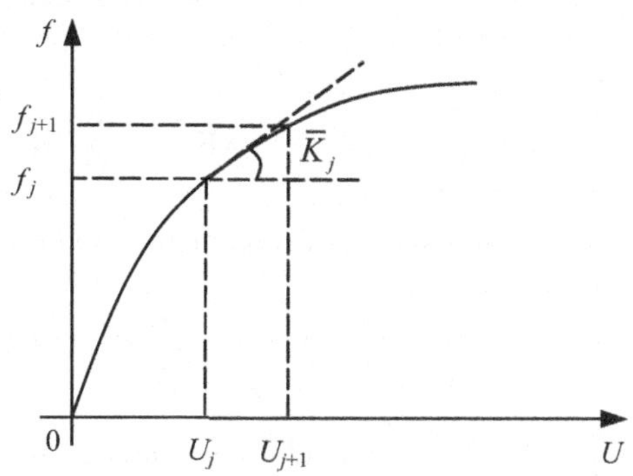

Fig. 5.4 The relationship between resorting force and displacement

The method can be extended to multi-degree of freedom system. Similar to Eq. (5.72), there exists

$$\mathbf{F}_{j+1} = \mathbf{F}_j + \bar{\mathbf{K}}_j \Delta\mathbf{U}_j \tag{5.73}$$

Substituting the above equation into Eq. (5.71), and subtracting Eq. (5.70), the incremental equation of non-linear structure can be written as follows:

$$\mathbf{M}\Delta\ddot{\mathbf{U}}_j + \mathbf{C}\Delta\dot{\mathbf{U}}_j + \bar{\mathbf{K}}_j\Delta\mathbf{U}_j = -\mathbf{M}\Delta\ddot{\mathbf{U}}_{g,j} \tag{5.74}$$

Using Eqs. (5.63) and (5.64), velocity increment and displacement increment can be expressed as follows:

$$\dot{\mathbf{U}}_j = \ddot{\mathbf{U}}_j\Delta t + \frac{1}{2}\ddot{\mathbf{U}}_j\Delta t \tag{5.75}$$

$$\mathbf{U}_j = \dot{\mathbf{U}}_j\Delta t + \frac{1}{2}\ddot{\mathbf{U}}_j\Delta t^2 + \frac{1}{6}\ddot{\mathbf{U}}_j\Delta t^2 \tag{5.76}$$

Considering Eq. (5.62), Eq. (5.76) can be rewritten as follows:

$$\Delta\ddot{\mathbf{U}}_j = 6\frac{\Delta\mathbf{U}_j}{\Delta t^2} - 6\frac{\Delta\dot{\mathbf{U}}_j}{\Delta t} - 3\ddot{\mathbf{U}}_j \tag{5.77}$$

Combining the above equation and Eq. (5.62), Eq. (5.75) can be rewritten as follows:

$$\Delta\dot{\mathbf{U}}_j = 3\frac{\Delta\mathbf{U}_j}{\Delta t} - 3\dot{\mathbf{U}}_j - \frac{1}{2}\ddot{\mathbf{U}}_j\Delta t \tag{5.78}$$

Substituting Eqs. (5.77) and (5.78) in Eq. (5.74), a pseudo-static incremental equation is established:

$$\Delta\mathbf{K}_j\Delta\mathbf{U}_j = \Delta\mathbf{P}_j \tag{5.79}$$

where

$$\Delta\mathbf{K}_j = \frac{6}{\Delta t^2}\mathbf{M} + \frac{3}{\Delta t}\mathbf{C} + \bar{\mathbf{K}}_j \tag{5.80}$$

is the pseudo-static stiffness matrix, and

$$\Delta\mathbf{P}_j = \mathbf{M}\left(-\Delta\ddot{\mathbf{U}}_{g,j} + \frac{6}{\Delta t}\dot{\mathbf{U}}_j + 3\ddot{\mathbf{U}}_j\right) + \mathbf{C}\left(3\dot{\mathbf{U}}_j + \frac{1}{2}\ddot{\mathbf{U}}_j\Delta t\right) \tag{5.81}$$

is the pseudo-static load incremental vector.

According to Eqs. (5.77–5.79), the structural response can be obtained step by step. However, as the pseudo-static load incremental vector is determined by $\dot{\mathbf{U}}_j$ and $\ddot{\mathbf{U}}_j$, which is given by the last time interval, the numerical error could be accumulated gradually. This will result in divergence of the result. In order to reduce the accumulation error, an acceleration modified algorithm should be adopted. In other words, the acceleration increment is not given by Eq. (5.77) but by the incremental dynamic equation, i.e., Eq. (5.74). That is:

$$\Delta\ddot{\mathbf{U}}_j = -\Delta\ddot{\mathbf{U}}_{g,j} - \mathbf{M}^{-1}\left(\mathbf{C}\Delta\dot{\mathbf{U}}_j + \bar{\mathbf{K}}_j\Delta\mathbf{U}_j\right) \tag{5.82}$$

Previous studies indicate that this acceleration modification can avoid numerical error accumulation to a certain degree. However, the algorithm does not help to improve the convergence and stability of the linear acceleration algorithm. In fact, algorithm convergence and stability problems exist in all time history integration algorithms. The algorithm convergence means the error will approach infinitesimal when time step approaches to infinitesimal. It can be proved that the linear acceleration algorithm is of two-order convergence, i.e. its truncation error is $O(\Delta t^3)$. Meanwhile, the algorithm stability means whether the solution increases infinitely or oscillates due to the initial condition or numerical error in computation. When the solution does not increase infinitely or oscillate and does not relate to time step, the algorithm is called unconditionally stable. Conversely, it is called conditionally stable. The linear acceleration algorithm is conditionally stable and the maximum time step is $0.55T_{\min}$, where $T_{\min}$ is the minimal structural period.

5.2.2 *Generalized α-Algorithm*

There is a lot of researches (Hilbert et al. 1977; Wod et al. 1981; Chung and Hulbert 1993) which focuse on the stability of numerical integration algorithm. Among those, the generalized α-algorithm (Chung and Hulbert 1993) is an excellent algorithm because of its suitability and unconditioned stability. Herein, an incremental algorithm is taken as example to illustrate the process of the generalized α-algorithm. In generalized α-algorithm, displacement, velocity and acceleration among the time interval (t_j, t_{j+1}) take the value of the linear combination of corresponding quantities at the time of t_j and t_{j+1}.

$$\mathbf{U}_{j+1-\alpha_f} = \left(1-\alpha_f\right)\mathbf{U}_{j+1} + \alpha_f\mathbf{U}_{\mathrm{j}} = \mathbf{U}_{\mathrm{j}} + \left(1-\alpha_f\right)\Delta\mathbf{U}_{\mathrm{j}} \tag{5.83}$$

$$\dot{\mathbf{U}}_{\mathrm{j}+1-\alpha_{\mathrm{f}}} = \left(1-\alpha_f\right)\dot{\mathbf{U}}_{\mathrm{j}+1} + \alpha_f\dot{\mathbf{U}}_{\mathrm{j}} = \dot{\mathbf{U}}_{\mathrm{j}} + \left(1-\alpha_f\right)\Delta\dot{\mathbf{U}}_{\mathrm{j}} \tag{5.84}$$

$$\ddot{\mathbf{U}}_{\mathrm{j}+1-\alpha_{\mathrm{m}}} = (1-\alpha_m)\ddot{\mathbf{U}}_{\mathrm{j}+1} + \alpha_m\ddot{\mathbf{U}}_{\mathrm{j}} = \ddot{\mathbf{U}}_{\mathrm{j}} + (1-\alpha_m)\Delta\ddot{\mathbf{U}}_{\mathrm{j}} \tag{5.85}$$

where α_f and α_m are adjustment parameters corresponding to a certain percentile. For example

$$t_{j+1-\alpha_f} = t_j + \left(1 - \alpha_f\right)\Delta t \tag{5.86}$$

Let the input of ground acceleration

$$\ddot{\mathbf{U}}_{g,j+1-\alpha_f} = \left(1 - \alpha_f\right)\ddot{\mathbf{U}}_{g,j+1} + \alpha\ddot{\mathbf{U}}_{g,j} = \ddot{\mathbf{U}}_{g,j} + \left(1 - \alpha_f\right)\Delta\ddot{\mathbf{U}}_{g,j} \tag{5.87}$$

For a linear system, substituting Eqs. (5.73–5.75) and (5.77) into Eq. (5.29) and subtracting the equilibrium equation at t_j, it can derive

$$\begin{aligned}(1 - \alpha_m)\mathbf{M}\Delta\ddot{\mathbf{U}}_j + \left(1 - \alpha_f\right)\mathbf{C}\Delta\dot{\mathbf{U}}_j + \left(1 - \alpha_f\right)\bar{\mathbf{K}}_j\Delta\mathbf{U}_j \\ = -\left(1 - \alpha_f\right)\mathbf{M}\Delta\ddot{\mathbf{U}}_{g,j}\end{aligned} \tag{5.88}$$

Meanwhile, a constant acceleration between t_j and t_{j+1} is assumed and takes the value of

$$\ddot{\mathbf{U}} = (1 - \gamma)\ddot{\mathbf{U}}_j + \ddot{\mathbf{U}}_{j+1} \tag{5.89}$$

An equivalent expression of the above equation is

$$\ddot{\mathbf{U}} = (1 - 2\beta)\ddot{\mathbf{U}}_j + 2\beta\ddot{\mathbf{U}}_{j+1} \tag{5.90}$$

After integration, the velocity and displacement at the time of t_{j+1} are

$$\dot{\mathbf{U}}_{j+1} = \dot{\mathbf{U}}_j + \ddot{\mathbf{U}}\Delta t \tag{5.91}$$

$$\mathbf{U}_{j+1} = \mathbf{U}_j + \dot{\mathbf{U}}_j\Delta t + \frac{1}{2}\ddot{\mathbf{U}}_j\Delta t^2 \tag{5.92}$$

Substituting Eq. (5.89) into (5.91) and Eq. (5.90) into (5.92), velocity and displacement increments become

$$\Delta\dot{\mathbf{U}}_j = (1 - \gamma)\ddot{\mathbf{U}}_j\Delta t + \gamma\ddot{\mathbf{U}}_{j+1}\Delta t = \ddot{\mathbf{U}}_j\Delta t + \gamma\Delta\ddot{\mathbf{U}}_j\Delta t \tag{5.93}$$

$$\Delta\mathbf{U}_j = \dot{\mathbf{U}}_j\Delta t + \left(\frac{1}{2} - \beta\right)\ddot{\mathbf{U}}_j\Delta t^2 + \beta\ddot{\mathbf{U}}_{j+1}\Delta t^2 = \dot{\mathbf{U}}_j\Delta t + \left(\frac{1}{2}\ddot{\mathbf{U}}_j + \beta\Delta\ddot{\mathbf{U}}_j\right)\Delta t^2 \tag{5.94}$$

Then the acceleration increment can be expressed as follows:

$$\Delta\ddot{\mathbf{U}}_j = \frac{1}{\beta}\left(\frac{\Delta\mathbf{U}_j}{\Delta t^2} - \frac{\dot{\mathbf{U}}_j}{\Delta t} - \frac{1}{2}\ddot{\mathbf{U}}_j\right) \tag{5.95}$$

Substituting above equation into Eq. (5.93) will give:

$$\Delta\dot{\mathbf{U}}_j = \frac{\gamma}{\beta}\left(\frac{\Delta\mathbf{U}_j}{\Delta t} - \dot{\mathbf{U}}_j\right) - \left(1 - \frac{\gamma}{2\beta}\right)\ddot{\mathbf{U}}_j\Delta t \tag{5.96}$$

Substituting Eqs. (5.95) and (5.96) into Eq. (5.88), a pseudo-static incremental equation is established.

$$\mathbf{K}^*\Delta\mathbf{U}_j = \Delta\mathbf{P}_j \tag{5.97}$$

where

$$\mathbf{K}^* = \frac{1-\alpha_m}{\beta\Delta t^2}\mathbf{M} + \frac{\gamma(1-\alpha_f)}{\beta\Delta t}\mathbf{C} + (1-\alpha_f)\mathbf{K} \tag{5.98}$$

is called the pseudo-static incremental stiffness.

$$\begin{aligned}\Delta\mathbf{P}_j = \mathbf{M}&\left[-(1-\alpha_f)\Delta\ddot{\mathbf{U}}_{g,j} + \frac{1-\alpha_m}{\beta\Delta t}\dot{\mathbf{U}}_j + \frac{1-\alpha_m}{2\beta}\ddot{\mathbf{U}}_j\right]\\ &+\mathbf{C}\left[\frac{\gamma(1-\alpha_f)}{\beta}\dot{\mathbf{U}}_j - (1-\alpha_f)\left(1-\frac{\gamma}{2\beta}\right)\ddot{\mathbf{U}}_j\Delta t\right]\end{aligned} \tag{5.99}$$

is called the pseudo-static incremental load.

When $\beta, \gamma, \alpha_f, \alpha_m$ are given, dynamic response of structure can be obtained using the above pseudo-static incremental equation. Meanwhile, if stiffness matrix in Eq. (5.98) is the incremental stiffness matrix $\bar{\mathbf{K}}_j$ in time interval (t_j, t_{j+1}), the Eq. (5.97) can be used to solve the non-linear dynamic response of structures.

It is worth to point out that the increment Eq. (5.88) does not satisfy the original dynamic equation strictly when $\alpha_m \neq \alpha_f$. However, this defect would lead to the unconditional stability of the algorithm. The reason lies in that the inconsistency compensates the deviation brought by the assumptions in Eqs. (5.87) and (5.88). In fact, according to different values of $\beta, \gamma, \alpha_f, \alpha_m$, the generalized α-algorithm could become into others different algorithms.

(1) When $\gamma = \frac{1}{2}, \beta = 0, \alpha_m = 0, \alpha_f = 0$, the generalized α-algorithm becomes into the central difference method which is conditionally stable.
(2) When $\gamma = \frac{1}{2}, \beta = \frac{1}{6}, \alpha_m = 0, \alpha_f = 0$, the generalized α-algorithm becomes into the linear acceleration algorithm which is conditionally stable.
(3) When $\gamma = \frac{1}{2}, \beta = \frac{1}{4}, \alpha_m = 0, \alpha_f = 0$, the generalized α-algorithm becomes into the constant acceleration algorithm which is unconditionally stable.
(4) When $\gamma = \frac{1}{2} + \alpha_f, \beta = \frac{1}{4}(1+\alpha_f)^2, \alpha_m = 0, \alpha_f = \frac{1-\rho}{1+\rho}$, the generalized α-algorithm becomes into the HHT-α algorithm (Hilbert et al. 1977) which is unconditionally stable.

(5) When $\gamma = \frac{1}{2} - \alpha_m, \beta = \frac{1}{4}(1 - \alpha_m)^2, \alpha_m = \frac{\rho - 1}{1 + \rho}, \alpha_f = 0$, thegeneralized α-algorithm becomes into the WBZ-α algorithm (Wod et al. 1981) which is unconditionally stable.

At last, when

$$\gamma = \frac{1}{2} - \alpha_m + \alpha_f \tag{5.100}$$

$$\beta = \frac{1}{4}(1 - \alpha_m + \alpha_f) \tag{5.101}$$

$$\alpha_m = \frac{2\rho - 1}{1 + \rho} \tag{5.102}$$

$$\alpha_f = \frac{\rho}{1 + \rho} \tag{5.103}$$

it is the general α-algorithm.

The HHT-α, WBZ-α and general α-algorithms are all unconditionally stable when $\rho < 1$. In fact, when $\rho = 1$, these three algorithms become into the constant acceleration algorithm which is unconditionally stable. However, the results from the constant acceleration algorithm only have one-order accuracy while those from the above three algorithms have two-order accuracy.

For linear structures, precious integration method (Zhong 2002) is another excellent algorithm, which could be learned in related literature.

5.3 Stochastic Seismic Response Analysis of Structures

Actually, a real engineering structure should be regarded as a stochastic structure because its mass, stiffness and damping parameters are all random variables in essence. Meanwhile, the ground motion is a typical stochastic process. Considering such background, the stochastic seismic response analysis method should be introduced. For general stochastic systems, a probability density evolution method (PDEM) proposed by Li and Chen (2009) provides a promising tool for this target.

5.3.1 Principle of Preservation of Probability

Consider an n-dimensional stochastic dynamical system

$$\dot{\mathbf{Y}} = \mathbf{A}(\mathbf{Y}, t), \mathbf{Y}(t_0) = \mathbf{Y}_0 \tag{5.104}$$

where $\mathbf{Y} = (Y_1, Y_2, \ldots, Y_n)^{\mathrm{T}}$ is the n-dimensional state vector, $\mathbf{Y}_0 = \left(Y_{0,1}, Y_{0,2}, \ldots, Y_{0,n}\right)^{\mathrm{T}}$ is the corresponding initial vector, $\mathbf{A}(\cdot)$ is a deterministic operator vector.

The state equation essentially establishes a mapping from $\mathbf{Y}_0$ to $\mathbf{Y}_t = \mathbf{Y}(t)$, which can be expressed as follows:

$$\mathbf{Y}_t = g(\mathbf{Y}_0, t) \tag{5.105}$$

where $g(\cdot)$ is mapping operator from $\mathbf{Y}_0$ to $\mathbf{Y}(t)$.

When $\mathbf{Y}_0$ is a random vector, $\{\mathbf{Y}_0 \in \Omega_0\}$ will be a random event. Herein, Ω_0 is an arbitrary domain in the distribution range of $\mathbf{Y}_0$. According to the stochastic state Eq. (5.105), $\mathbf{Y}_0$ will be changed to $\mathbf{Y}_t$ at time t. The domain Ω_0 to which $\mathbf{Y}_0$ belongs at t_0 is accordingly changed to Ω_t to which $\mathbf{Y}_t$ belongs at t (Fig. 5.5), i.e.

$$\Omega_t = g(\Omega_0, t) \tag{5.106}$$

Because the probability is preserved in the mapping of any arbitrary element event, we have

$$\int\limits_{\Omega_t} p_{\mathbf{Y}}(\mathbf{y}, t)d\mathbf{y} = \int\limits_{\Omega_{t_0}} p_{\mathbf{Y}}(\mathbf{y}, t_0)d\mathbf{y} \tag{5.107}$$

where $p_{\mathbf{Y}}(\mathbf{y}, t)$ is the joint probability density function of $\mathbf{Y}_t$.

Evidently, the above equation is also correct at $t + \Delta t$, which will give:

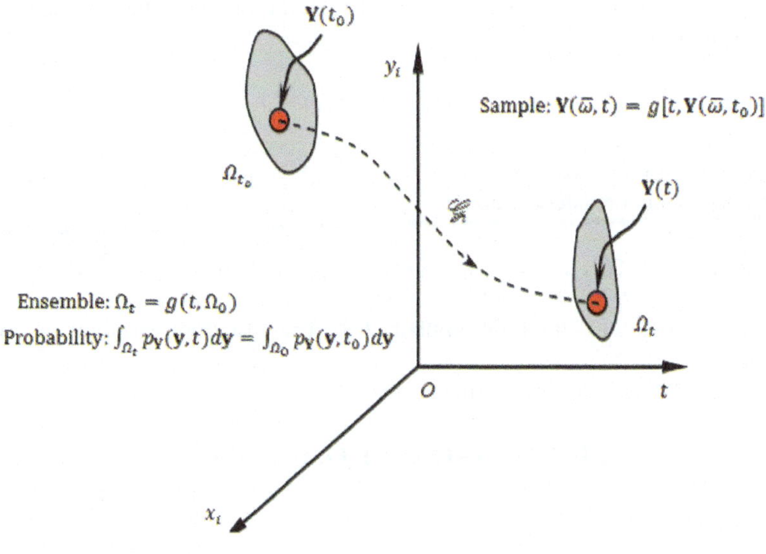

Fig. 5.5 Dynamical system, mapping and probability evolution ($\mathbf{Y}_t = \mathbf{Y}(t)$, $\mathbf{Y}_o = \mathbf{Y}(t_0)$)

$$\frac{\mathrm{D}}{\mathrm{D}t}\int\limits_{\Omega_t} p_{\mathbf{Y}}(\mathbf{y},t)d\mathbf{y} = 0 \tag{5.108}$$

where $\mathrm{D}(\cdot)/\mathrm{D}t$ denotes the total derivative.

Equation (5.108) actually indicates the principle of preservation of probability in a stochastic dynamical system. Since it is the result from the perspective that the probability of a random event is invariant, we refer to it as the random event description of the principle of preservation of probability (Li and Chen 2008).

5.3.2 The Generalized Probability Density Evolution Equation

Without loss of generality, let us consider the equation of motion of a multi-degree-of-freedom (MDOF) system

$$\mathbf{M}(\boldsymbol{\eta})\ddot{\mathbf{U}} + \mathbf{C}(\boldsymbol{\eta})\dot{\mathbf{U}} + \mathbf{f}(\boldsymbol{\eta},\mathbf{U}) = \boldsymbol{\Gamma}\xi(t) \tag{5.109}$$

where $\boldsymbol{\eta} = \left(\eta_1, \eta_1, \ldots, \eta_{s_1}\right)$ is the random parameters involved in the system. If the excitation is a stochastic ground acceleration $\xi(t) = \ddot{X}_g(t)$, then $\boldsymbol{\Gamma} = -\mathbf{M}1$, $1 = (1, 1, \ldots, 1)^{\mathrm{T}}$. Here, $\ddot{\mathbf{U}}, \dot{\mathbf{U}}, \mathbf{U}$ are the accelerations, velocities and displacements of the system relative to the ground. $\mathbf{M}$, $\mathbf{C}$ are the mass and damping matrices of the system, respectively, and $\mathbf{f}(\cdot)$ is the restoring force vector.

In Chap. 3, we know that the stochastic ground motion can be modeled as a physical stochastic process, which means the excitation could be represented by a random function

$$\xi(t) = \ddot{U}_g(\boldsymbol{\zeta}, t) \tag{5.110}$$

where $\boldsymbol{\zeta} = \left(\zeta_1, \zeta_2, \ldots, \zeta_{s_2}\right)$.

For notational consistency, denote

$$\boldsymbol{\Theta} = (\boldsymbol{\eta}, \boldsymbol{\zeta}) = \left(\eta_1, \eta_1, \ldots, \eta_{s_1}, \zeta_1, \zeta_2, \ldots, \zeta_{s_2}\right) = (\Theta_1, \Theta_2, \ldots, \Theta_s) \tag{5.111}$$

in which $s = s_1 + s_2$ is the total number of the basic random variables involved in the system.

Then Eq. (5.109) can be rewritten into:

$$\mathbf{M}(\boldsymbol{\Theta})\ddot{\mathbf{U}} + \mathbf{C}(\boldsymbol{\Theta})\dot{\mathbf{U}} + \mathbf{f}(\boldsymbol{\Theta},\mathbf{U}) = \mathbf{F}(\boldsymbol{\Theta}, t) \tag{5.112}$$

where $\mathbf{F}(\boldsymbol{\Theta}, t) = \boldsymbol{\Gamma}\ddot{U}_g(\zeta, t)$

Equation (5.112) is a basic equation to be resolved in which all the randomness from the initial conditions, excitations and system parameters is involved and exposed in a unified manner.

If we are interested in other physical quantities $\mathbf{Z} = (Z_1, Z_2, \ldots, Z_m)^{\mathrm{T}}$ in the system (e.g. the stress, internal forces, etc.), then the augmented system $(\mathbf{Z}, \boldsymbol{\Theta})$ is probability preserved because all the random factors are involved, thus according to Eq. (5.108), that is

$$\frac{\mathrm{D}}{\mathrm{D}t} \int\limits_{\Omega_t \times \Omega_\theta} p_{\mathbf{Z}\Theta}(\mathbf{z}, \boldsymbol{\theta}, t) d\mathbf{z} d\boldsymbol{\theta} = 0 \tag{5.113}$$

where $\Omega_t \times \Omega_\theta$ is an arbitrary domain in the augmented state space $\Omega \times \Omega_\theta$, Ω_θ is the distribution range of the random vector $\boldsymbol{\Theta}$, $p_{\mathbf{Z}\Theta}(\mathbf{z}, \boldsymbol{\theta}, t)$ is the joint probability density function of $(\mathbf{Z}(t), \boldsymbol{\Theta})$.

After a series of mathematical manipulations, the following expression will be derived (Li and Chen 2009):

$$\int\limits_{\Omega_0 \times \Omega_\theta} \left(\frac{\partial p_{\mathbf{Z}\boldsymbol{\Theta}}(\mathbf{z}, \boldsymbol{\theta}, t)}{\partial t} + \sum_{j=1}^{m} \dot{Z}_j(\boldsymbol{\theta}, t) \frac{\partial p_{\mathbf{Z}\boldsymbol{\Theta}}(\mathbf{z}, \boldsymbol{\theta}, t)}{\partial z_j} \right) d\mathbf{z} d\boldsymbol{\theta} = 0 \tag{5.114}$$

which holds for an arbitrary $\Omega_0 \times \Omega_\theta \in \Omega \times \Omega_{\boldsymbol{\Theta}}$.

Then we have:

$$\frac{\partial p_{\mathbf{Z}\boldsymbol{\Theta}}(\mathbf{z}, \boldsymbol{\theta}, t)}{\partial t} + \sum_{j=1}^{m} \dot{Z}_j(\boldsymbol{\theta}, t) \frac{\partial p_{\mathbf{Z}\boldsymbol{\Theta}}(\mathbf{z}, \boldsymbol{\theta}, t)}{\partial z_j} = 0 \tag{5.115}$$

Specifically, when $m = 1$, Eq. (5.115) becomes:

$$\frac{\partial p_{Z\boldsymbol{\Theta}}(z, \boldsymbol{\theta}, t)}{\partial t} + \dot{Z}(\boldsymbol{\theta}, t) \frac{\partial p_{Z\boldsymbol{\Theta}}(z, \boldsymbol{\theta}, t)}{\partial z} = 0 \tag{5.116}$$

which is a one-dimensional partial differential equation.

Equations (5.115) and (5.116) are referred to as generalized probability density evolution equations (Li and Chen 2008). They reveal the intrinsic connections between a stochastic dynamical system and its deterministic counterpart. It is remarkable that the dimension of the general probability density evolution equation is free from the dimension (or the degree-of-freedom) of the original system. This distinguishes the generalized probability density evolution equation from the traditional probability density evolution equations (e.g. Liouville, Dostupov-Pugachev and FPK equations), in which the dimension must be identical to the dimension of the original state equation (twice of the degree-of-freedom).

Particularly, if the physical quantity Z of interest is the displacement X of the system, Eq. (5.116) becomes:

$$\frac{\partial p_{U\Theta}(u, \boldsymbol{\theta}, t)}{\partial t} + \dot{U}(\boldsymbol{\theta}, t)\frac{\partial p_{U\Theta}(u, \boldsymbol{\theta}, t)}{\partial u} = 0 \tag{5.117}$$

The equation can be changed as:

$$\frac{\partial p_{U\Theta}(u, \boldsymbol{\theta}, t)}{\partial t} = -\dot{U}(\boldsymbol{\theta}, t)\frac{\partial p_{U\Theta}(u, \boldsymbol{\theta}, t)}{\partial u} \tag{5.118}$$

Then the rule is clearly revealed by the general probability density evolution equation: in the evolution of a general dynamical system, the time-variant rate of the joint probability density function of displacement and source random parameters is proportional to the space-variant rate, and the coefficient is the instantaneous velocity. In other words, the change of probability density is decided by the change of physical states. This demonstrates evidently that, in a stochastic system, the evolution of probability density is not disordered, but admits the restrict physical law. Clearly, this holds for general physical systems with underlying randomness.

Although in principle, the generalized probability density evolution equation holds for any arbitrary dimension, in most cases one or two-dimensional probability density evolution equations are concerned. For simplicity and clarity, in the following sections, we will focus on the one-dimensional probability density evolution equation.

The boundary condition for Eq. (5.116) is

$$p_{Z\Theta}(z, \boldsymbol{\theta}, t)\big|_{z\to\pm\infty} = 0 \tag{5.119}$$

while the initial condition is:

$$p_{Z\Theta}(z, \boldsymbol{\theta}, t)\big|_{t\to t_0} = \delta(z - z_0)p_{\Theta}(\boldsymbol{\theta}) \tag{5.120}$$

where z_0 is the deterministic initial value.

Solving the Eq. (5.116) with (5.119) and (5.120), the instantaneous probability density function of $Z(t)$ can be obtained by

$$p_Z(z, t) = \int_{\Omega_\theta} p_{Z\Theta}(z, \boldsymbol{\theta}, t)d\boldsymbol{\theta} \tag{5.121}$$

5.3.3 *Numerical Method for Solving General Probability Density Evolution Equation*

Probability density evolution, i.e., Eq. (5.116), can be analytically solved only for a very simple system. For general structural system, the practical choice to solve

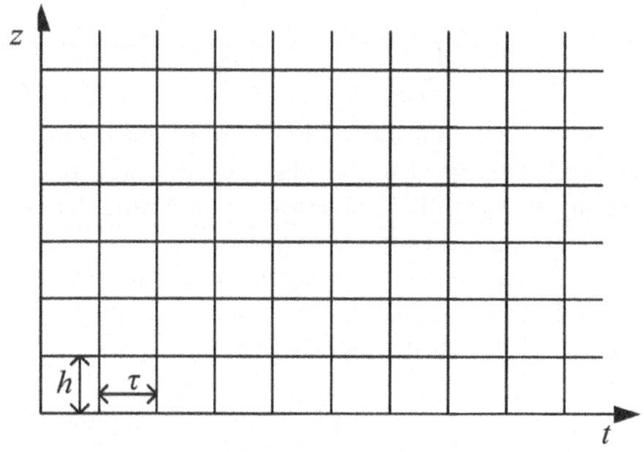

Fig. 5.6 Difference meshes division

the equation resorts to a numerical method. The procedure of the numerical method includes the following three steps:

(1) Discretize random space $\boldsymbol{\Omega}$ into a series of subspace Ω_G.
(2) For each subspace, solve deterministic dynamic equation, i.e., Eq. (5.112) to give $\dot{\mathbf{Z}}(\theta_q)$.
(3) To solve probability density evolution, i.e., Eq. (5.116) with (5.119) and (5.120).
(4) Taking integration with respect to $\boldsymbol{\Theta}_\varsigma$ to give the probability density of Z.

In the above procedure, the steps (1) and (4) are simple, and the step (2) is a deterministic seismic analysis problem which has been introduced in Sect. 5.2 in detail. Therefore, the key of the above procedure is step (3).

For the partial differential equation with an initial value, the finite difference method is an excellent numerical analysis method (Li and Chen 2009). In fact, for special $\mathbf{Z}_\varsigma$, Eq. (5.116) is a partial differential equation of hyperbolic type with various coefficients. In order to construct finite difference approximate equations, the computation region should be covered by the meshes formed by two lines which are parallel to x and t axis, respectively. The meshes are shown in Fig. 5.6 and the length of meshes is $\Delta z = h$ and $\Delta t = \tau$, where h and τ are the step length of space and time, respectively. The intersections of meshes are called mesh nodes. The finite difference method is a numerical method which replaces derivation with difference quotient, constructs approximate difference equations and solves the equations at mesh nodes.

Taking two-variable function $u(z, t)$ as example, there are three approximate expressions for the one-order partial derivative of u with respect to z at the point (z_m, t_n).

$$\left(\frac{\partial u}{\partial z}\right)_m^n \approx \frac{u(z_{m+1}, t_n) - u(z_m, t_n)}{h} = \frac{u_{m+1}^n - u_m^n}{h} \tag{5.122}$$

$$\left(\frac{\partial u}{\partial z}\right)_m^n \approx \frac{u(z_m, t_n) - u(z_{m-1}, t_n)}{h} = \frac{u_m^n - u_{m-1}^n}{h} \tag{5.123}$$

$$\left(\frac{\partial u}{\partial z}\right)_m^n \approx \frac{u(z_{m+1}, t_n) - u(z_{m-1}, t_n)}{2h} = \frac{u_{m+1}^n - u_{m-1}^n}{2h} \tag{5.124}$$

Above equations are called forward, backward and center difference schemes of $u(z, t)$ with respect to z at the point (z_m, t_n), respectively. Using the Taylor series expansion, the truncation errors of the former two expressions are $o(h)$ while that of the latest one is $o(h^2)$.

Using the above difference expressions, different difference equations and solution schemes can be deduced. For example, when the forward difference scheme is applied to Eq. (5.116), it can give:

$$\frac{p_m^{n+1} - p_m^n}{\tau} + \dot{z}_{l,n}\frac{p_{m+1}^n - p_m^n}{h} = 0 \tag{5.125}$$

It can be rewritten as follows:

$$p_m^{n+1} = \left(1 + r\dot{z}_{l,n}\right)p_m^n - r\dot{z}_{l,n}p_{m+1}^n \tag{5.126}$$

where $r = \tau/h$ is called the mesh ratio.

Then, just as Fig. 5.7 indicates, p_m^{n+1} at the node $(m, n + 1)$ which is located in $n+1$ time layer can be determined by p_m^n and p_{m+1}^n at two neighbor nodes (m, n) and $(m + 1, n)$ which are both located in the nth time layer. For a structure with special z_ς and being still at the beginning, considering initial condition (5.119), p_m^n at the nodes can be given step by step according to Eq. (5.127) when p_j^0 takes the value of $\delta_{0j}p_\varsigma\left(z_\varsigma\right)$. Herein, δ is a Kroneber sign. As the above difference scheme only contains one unknown variable, it is called explicit difference scheme.

A detailed study indicates that the simple forward or backward difference scheme owns only one-order accuracy. Moreover, the forward difference scheme is convergent and stable only if $\dot{\mathbf{z}}_{l,n} < 0$. Apparently, it is not suitable to solve Eq. (5.116). Similarly, the backward difference scheme is convergent and stable only if $\dot{\mathbf{z}}_{l,n} > 0$.

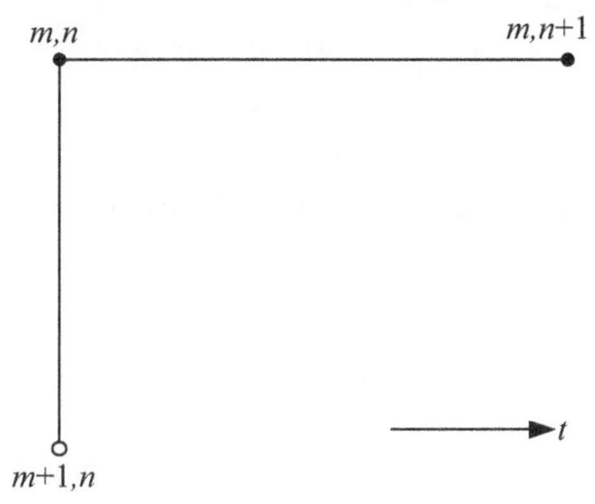

Fig. 5.7 One-side difference scheme

Meanwhile, the center difference scheme is always unstable. Thus, these classical difference schemes are not suitable to solve Eq. (5.116).

Fortunately, Lax-Wendroff difference scheme is suitable for solving Eq. (5.116) (Li and Chen 2003) and it owns two-order accuracy. The Lax-Wendroff difference scheme can be derived as follows.

The Taylor series expansion of $p(x, t)$ can be written as follows:

$$p_m^{n+1} = p_m^n + \tau\left(\frac{\partial p}{\partial t}\right)_m^n + \frac{\tau^2}{2}\left(\frac{\partial^2 p}{\partial t^2}\right)_m^n + O(\tau^3) \tag{5.127}$$

According to Eq. (5.116), there is

$$\frac{\partial p}{\partial t} = -\dot{\mathbf{z}}_l \frac{\partial p}{\partial x_l} \tag{5.128}$$

$$\frac{\partial^2 p}{\partial t^2} = \frac{\partial}{\partial t}\left(-\dot{\mathbf{z}}_l \frac{\partial p}{\partial x_l}\right) = \dot{\mathbf{z}}_l^2 \frac{\partial^2 p}{\partial x_l^2} \tag{5.129}$$

Using center difference scheme, it can derive:

$$\left(\frac{\partial p}{\partial t}\right)_m^n = \frac{p_{m+1}^n - p_{m-1}^n}{2h} + O(h^2) \tag{5.130}$$

$$\left(\frac{\partial^2 p}{\partial t^2}\right)_m^n = \frac{p_{m+1}^n - 2p_m^n + p_{m-1}^n}{h^2} + O(h^2) \tag{5.131}$$

Substituting the above four equations in Eq. (5.128) and neglecting high-order terms, the Lax-Wendroff difference scheme is established.

$$\begin{aligned} p_{m+1}^n &= p_m^n - \frac{r}{2}\dot{\mathbf{z}}_{l,n}\left(p_{m+1}^n - p_{m-1}^n\right) + \frac{r^2}{2}\dot{\mathbf{z}}_{l,n}^2\left(p_{m+1}^n - 2p_m^n + p_{m-1}^n\right) \\ &= \frac{1}{2}\left(r^2\dot{\mathbf{z}}_{l,n}^2 - r\dot{\mathbf{z}}_{l,n}\right)p_{m+1}^n + \left(1 - r^2\dot{\mathbf{z}}_{l,n}^2\right)p_m^n + \frac{1}{2}\left(r^2\dot{\mathbf{z}}_{l,n}^2 + r\dot{z}_{l,n}\right)p_{m-1}^n \end{aligned} \tag{5.132}$$

where $r = \tau/h$.

The Lax-Wendroff difference scheme is shown in Fig. 5.8. Apparently, the Lax-Wendroff difference scheme is an explicit difference scheme. When the initial condition is given, using the value of p at three neighbor nodes $(m - 1, n)$, (m, n) and $(m + 1, n)$ in time layer n, p_m^n at all nodes in computation region can be calculated step by step.

The stability condition of the Lax-Wendroff difference scheme is (Li and Chen 2009):

$$r^2\dot{\mathbf{z}}_{l,n}^2 \le 1 \tag{5.133}$$

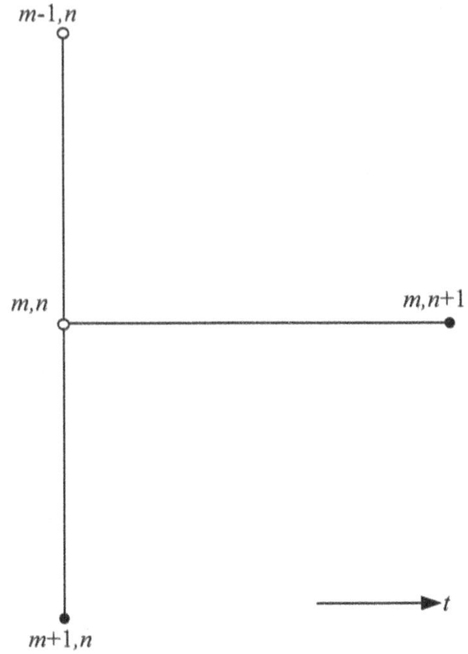

Fig. 5.8 Lax-Wendroff difference scheme

or

$$\left|r\dot{\mathbf{z}}_{l,n}\right| \leq 1 \tag{5.134}$$

Then, the value of r can be determined. In practice, $\left|\dot{\mathbf{z}}_{l,m}\right|_{\max}$ can be estimated preciously and r takes the value of

$$r = \frac{r_0}{\left|\dot{\mathbf{Z}}_{l,m}\right|_{\max}} \tag{5.135}$$

where $0 \leq r_0 \leq 1$ is an adjustment factor.

5.4 Seismic Reliability Analysis of Structures

Taking the dynamic equation and the general probability density evolution equation as a set of simultaneous equations, and introducing failure criteria of structures at different levels, the global reliability of structures under seismic ground motions can be solved (Li 2018).

To do so, introducing the screening operator firstly,

$$\mathcal{H}[f(\boldsymbol{u}(\boldsymbol{\theta},t))] = \begin{cases} 1 & f(\boldsymbol{u}(\boldsymbol{\theta},t)) \in \Omega_f \\ 0 & f(\boldsymbol{u}(\boldsymbol{\theta},t)) \in \Omega_s \end{cases} \tag{5.136}$$

where $\boldsymbol{u}$ is structural displacement response vector, $f(\cdot)$ is a function of $\boldsymbol{u}$, of which the form is defined by different failure criteria.

Then, taking a position of the structures as an observation window, and taking the structural displacement of the position, u_p, as an observation variable, when the structural state meets the failure criteria, the structure approaches failure state and therefore the probability carried by the observation variable will be dissipated. That is, when $\mathcal{H}[f(\boldsymbol{u}(\boldsymbol{\theta},t))] = 1$,

$$p_{U_p\boldsymbol{\Theta}}(u_p, \boldsymbol{\theta}, \tau) = 0 \tag{5.137}$$

For a probability dissipation system, at any instant of time, there exists

$$\Pr\{[U_p(t+dt), \boldsymbol{\theta}] \in \Omega_{t+dt} \times \Omega_{\boldsymbol{\Theta}}\} - \Pr\{[U_p(t), \boldsymbol{\theta}] \in \Omega_t \times \Omega_{\boldsymbol{\Theta}}\} = \delta p \tag{5.138}$$

where δp is the dissipation probability during dt interval

$$\delta p = \mathcal{H}[f(u(\boldsymbol{\theta},t))] \int_{\Omega_t \times \Omega_{\boldsymbol{\Theta}}} p_{U_p\boldsymbol{\Theta}}(u_p, \boldsymbol{\theta}, \tau) du_p \mathrm{d}\boldsymbol{\theta} \tag{5.139}$$

According to Eq. (5.138), the following equation can be deduced:

$$\frac{\partial p_{U_p\boldsymbol{\Theta}}(u_p, \boldsymbol{\theta}, t)}{\partial t} + \dot{u}_p(\boldsymbol{\theta}, t) \frac{\partial p_{U_p\boldsymbol{\Theta}}(u_p, \boldsymbol{\theta}, t)}{\partial u_p} = -\mathcal{H}[f(\boldsymbol{u}(\boldsymbol{\theta},t))] p_{U_p\boldsymbol{\Theta}}(u_p, \boldsymbol{\theta}, t) \tag{5.140}$$

The equation has a zero solution at $\mathcal{H} = 1$, and a non-zero solution $p_{U_p\boldsymbol{\Theta}}(u_p, \boldsymbol{\theta}, t)$ at $\mathcal{H} = 0$.

By combining the above equation with the dynamic analysis equation of structures, a set of equations for solving the reliability of the structure can be given as follows:

$$\begin{cases} \boldsymbol{M}\ddot{\boldsymbol{U}}(\boldsymbol{\Theta}_M) + \boldsymbol{C}\dot{\boldsymbol{U}}(\boldsymbol{\Theta}_C) + \boldsymbol{K}\boldsymbol{U}(\boldsymbol{\Theta}_K) = \boldsymbol{F}(\boldsymbol{\Theta}_F, t) \\ \frac{\partial p_{U_p\boldsymbol{\Theta}}(u_p,\boldsymbol{\theta},t)}{\partial t} + \dot{u}_p(\boldsymbol{\theta}, t) \frac{\partial p_{U_p\boldsymbol{\Theta}}(u_p,\boldsymbol{\theta},t)}{\partial u_p} = -\mathcal{H}[f(\boldsymbol{u}(\boldsymbol{\theta},t))] p_{U_p\boldsymbol{\Theta}}(u_p, \boldsymbol{\theta}, t) \\ p_{U_p\boldsymbol{\Theta}}(u_p, \boldsymbol{\theta}, t_0) = \delta(u_p - u_{p_0}) p_{\boldsymbol{\Theta}}(\boldsymbol{\theta}) \end{cases} \tag{5.141}$$

where $\ddot{\mathbf{U}}, \dot{\boldsymbol{U}}, \boldsymbol{U}$ are the acceleration, velocity, and displacement of the structures, respectively; $\boldsymbol{\Theta}_M, \boldsymbol{\Theta}_C, \boldsymbol{\Theta}_K, \boldsymbol{\Theta}_F$ are the physical parameters that reflect structural mass, damping, stiffness and dynamic excitation, respectively; $\boldsymbol{\Theta} = (\boldsymbol{\Theta}_M, \boldsymbol{\Theta}_C, \boldsymbol{\Theta}_K, \boldsymbol{\Theta}_F)$. The remaining symbols are as described in previous sections.

Solve Eq. (5.141), the joint probability distribution density $p_{U_p\boldsymbol{\Theta}}(u_p, \boldsymbol{\theta}, t)$ can be given, and

$$p_{U_p}(u_p, t) = \int_{\Omega_{\Theta}} p_{U_p\Theta}(u_p, \boldsymbol{\theta}, t)\mathrm{d}\boldsymbol{\theta} \tag{5.142}$$

While the global reliability of the structure is

$$P_s(t) = \int_{-\infty}^{+\infty} p_{U_p}(u_p, t)\mathrm{d}u_p \tag{5.143}$$

It is seen that Eq. (5.141) combine the basic physical equation, the law of randomness propagation in the physical system and the criteria that reflect the physical characteristics of structural failure. Therefore, the method is called the physical synthesis method, from which the global reliability of structures can be readily solved (Li 2018).

The physical synthesis method can solve the global reliability of general structures corresponding to various structural failure criteria. In general, structural failure criteria can be defined in the following ways:

(1) Failure at the material level;
(2) Failure at the components level of structures;
(3) Structural deformation exceeds the prescribed limits;
(4) Structural instability (collapse).

For the material failure, component destruction and structural deformation failure, the screening operator in Eq. (5.141) can be defined as in a uniform as follows:

$$\mathcal{H}[f(\boldsymbol{u}(\boldsymbol{\theta},t))] = \begin{cases} 1 & \bigcup_{i=1}^{n} g_i(R_i, S_i) \le 0 \\ 0 & otherwise \end{cases} \tag{5.144}$$

where $g_i(\cdot)$ is a functional function, R, S is the structural resistance and structural response, respectively; n is the number of possible failure modes.

For structural collapse, the screening operator for analyzing the global reliability of the structure is:

$$\mathcal{H}[S(\boldsymbol{u}, t)] = \begin{cases} 1 \; S(\boldsymbol{u}, t) \le 0 \\ 0 \; S(\boldsymbol{u}, t) > 0 \end{cases} \tag{5.145}$$

Here, the dynamic stability function S is defined as (Xu and Li 2015; Zhou and Li 2017)

$$S(\boldsymbol{u}, t) = E_{\text{eff - inp}}(\boldsymbol{u}, t) - E_{\text{eff - intr}}(\boldsymbol{u}, t) \tag{5.146}$$

where $\boldsymbol{u}$ is the structural displacement vector, t is time, $E_{eff-inp}$ is the effective work of external force, i.e.,

$$E_{\text{eff - inp}}(\boldsymbol{u}, t) = \int_0^t \mathbf{F}^T(t)\mathrm{d}\boldsymbol{u}(t) - \int_0^t \dot{\boldsymbol{u}}^T \mathbf{C}\dot{\boldsymbol{u}}\mathrm{d}t - \int_0^t \left(\int_V \boldsymbol{\sigma} : \dot{\boldsymbol{\varepsilon}}_p \mathrm{d}V \right) \mathrm{d}t \quad (5.147)$$

and $E_{\text{eff - intr}}$ is the effective intrinsic energy, i.e.,

$$E_{\text{eff - intr}}(\boldsymbol{u}, t) = \left| \mathbf{f}^T(\boldsymbol{u}, t)\boldsymbol{u} - \int_V \boldsymbol{\sigma} : \dot{\boldsymbol{\varepsilon}}_e \mathrm{d}V \right| \quad (5.148)$$

where F is the external load vector, $\dot{\boldsymbol{u}}$ is the structural velocity vector, C is the structural damping matrix; $\boldsymbol{\sigma}$ is the stress tensor, $\dot{\boldsymbol{\varepsilon}}_p$ is the plastic strain rate vector, f is the restoring force vector and $\dot{\boldsymbol{\varepsilon}}_e$ is the elastic strain rate vector.

References

Chung, J., Hulbert, G. M. (1993). A time integration algorithm for structural dynamics with improved numerical dissipation: The generalized-α method. *Journal of Applied Mechanics*, 60: 371-375.

Hilbert, H. M., Hughes, T. J. R. Taylor, R. L. (1977). Improved numerical dissipation for time integration algorithms in structural dynamics. *Earthquake Engineering & Structural Dynamics*, 5(3):283-292.

Li, J. (1997). Disaster response analysis and system control of composite engineering system. *Journal of Natural Disaster*, 6(3): 1–9 (in Chinese).

Li, J., & Chen, J. B. (2009). *Stochastic dynamics of structures*. Wiley.

Li, J., & Chen, J. B. (2008). The principle of preservation of probability and the generalized density evolution equation. *Structural Safety, 30*(1), 65–77.

Li, J. (2018). Advances in global reliability analysis of engineering structures. *China Civil Engineering Journal, 51*(8), 1–10.

Li, J., Chen, J. B. (2003). The probability density evolution method for analysis of dynamic nonlinear response of stochastic structures, *Acta Mechanica Sinica*, 35(6):716–722 (in Chinese).

Wood W L , Bossak M , Zienkiewicz O C, (1980) An alpha modification of Newmark's method. *International Journal for Numerical Methods in Engineering*, 15 (10):1562–1566

Xu, J., & Li, J. (2015). An energetic criterion for dynamic instability of structures under arbitrary excitations. *International Journal of Structural Stability and Dynamics, 15*(2), 1–32.

Zhong, W. X. (2002). Dual system of Applied Mechanics, Beijing: *Science Press* (in Chinese).

Zhou, H., & Li, J. (2017). Effective energy criterion for collapse of deteriorating structural systems. *Journal of Engineering Mechanics, 143*(12), 04017135.

Chapter 6
Seismic Reliability Analysis of Engineering Network (I)—Connectivity Reliability

6.1 Introduction

A common characteristic of lifeline engineering systems is that they usually cover a large area as a network which consists of many structures, such as industrial buildings, pipes and faculties. Therefore, besides the seismic performance and reliability evaluation of structures, the seismic performance and reliability of the entire network should be investigated.

The seismic reliability evaluation of a lifeline engineering system includes three aspects:

1. Reliability assessment of system elements or structures, which refers to identify the reliability of the nodes and the edges of networks.
2. Connectivity reliability analysis, which means to identify the connectivity reliability of node-pairs in networks.
3. Functional reliability analysis, which means to identify the probability of satisfying the demand of customers.

Obviously, structural reliability, which has been described in detail in Chaps. 4 and 5, is the basis to evaluate the network reliability, while the connectivity reliability and the functional reliability are two basic indexes for different assessment goals. In practice, a suitable reliability index should be selected considering application scenarios. For example, for the urban gas supply systems, which cannot work with leakages, only network connectivity reliability is needed to investigate. However, for the water distribution networks, when the purpose of evaluation is the estimation of damage state of the network, the connectivity reliability could be performed. On the other side, if the evaluation changes to assess whether citizens can be supplied with enough water after earthquakes, the functional reliability of the network should be investigated. Moreover, for the transportation systems, the network connectivity analysis is a more practical choice. For the power electricity systems, if a relationship between the equipment damage state and the voltage loss can be established, the

J. Li and W. Liu, *Lifeline Engineering Systems*,
https://doi.org/10.1007/978-981-15-9101-3_6

functional reliability of network can be performed. Otherwise, the connectivity reliability analysis should be identified. Obviously, incomparision with the functional reliability index, the connectivity reliability index has a broader application scope for general lifeline systems. Therefore, the connectivity reliability analysis of network is considered firstly while the functional reliability analysis that mainly used for water distribution networks will be introduced in the next chapter.

6.2 Foundation of System Reliability Analysis

The network connectivity reliability analysis is based on the principles of the graph theory and the probability theory. Herein, the related concepts of graph theory are introduced briefly.

6.2.1 Basic Concepts of Graph Theory

A graph is composed of a series of nodes and edges. Nodes connect with each other by edges. Generally, a graph can be expressed as the following set.

$$G = \{V, E\} \tag{6.1}$$

where the non-empty set $V = \{v_1, v_2, \ldots, v_n\}$ is the node set, and $E = \{e_1, e_2, \ldots, e_m\}$ is the edge set, in which the arbitrary edge $e_i = \{v_s, v_t\}$ constructs a binary relationship over the set G.

An edge is a directed edge when it points from one node to another node. Otherwise, it is an undirected edge. If all the edges in the edge set are directed edges, the graph is called a directed graph. However, if all the edges in the edge set are undirected edges, the graph is called an undirected graph. Moreover, the graph which contains both directed edges and undirected edges is called a mixed graph. For example, a bridge network shown in Fig. 6.1 is a mixed graph. The node set is V = {1, 2, 3, 4} and the edge set is E = {a, b, c, d, e}. Also, the node 1 is a source node and the node 2 is a sink node.

Fig. 6.1 A bridge network

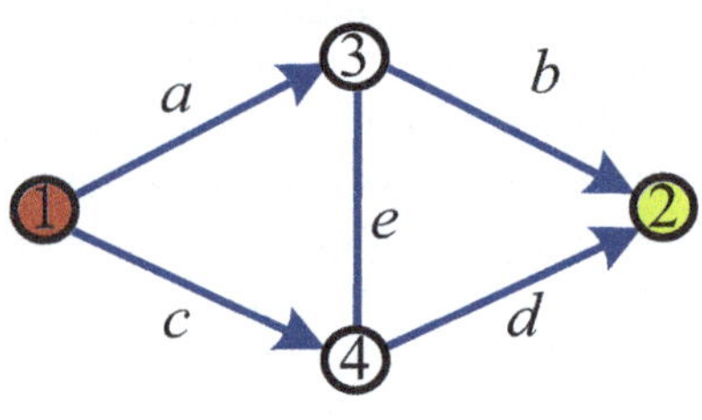

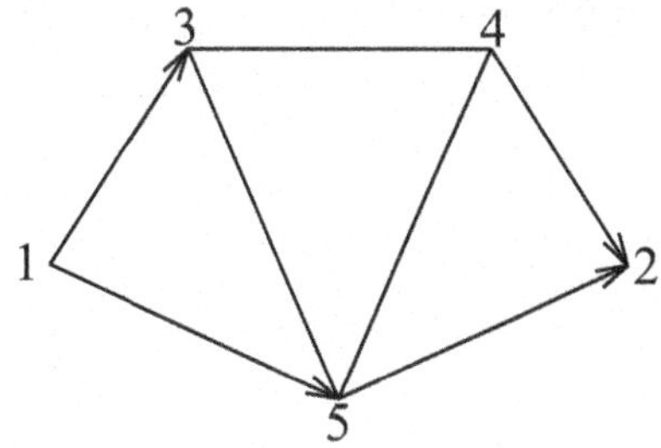

Fig. 6.2 A mixed graph

In this book, it is assumed that one edge just owns two different nodes, i.e. no edge has only one node and no edge connects more than two nodes.

The number of edges entering a node is called its in-degree while that of edges getting out of a node is called its out-degree. The degree of a node is the sum of its in-degree and out-degree. Apparently, the number of nodes, the maximal nodal degree and the number of edges indicate the complexity of a network. Herein, when calculating nodal degree, undirected edges should be viewed as two edges.

A path between nodes v_1 and v_2 is an edge set along which the matter can travel from v_1 to v_2. When any edge is removed from a path, it is no longer a path, then the path is called a minimal path which is usually represented by A. The number of edges in a minimal path is called the length of the path. As a minimal path does not consist of two same nodes or same edges, a minimal path with maximal length can only consist of no more than n nodes for a graph with n nodes. In other words, the maximal length of a minimal path is $n - 1$.

A mixed graph is shown in Fig. 6.2. Apparently, seven minimal paths exist between the source 1 and the sink 2. They are:

$A_1 : 1 \rightarrow 5 \rightarrow 2$ with a length of 2,
$A_2 : 1 \rightarrow 5 \rightarrow 4 \rightarrow 2$ with a length of 3,
$A_3 : 1 \rightarrow 3 \rightarrow 5 \rightarrow 2$ with a length of 3,
$A_4 : 1 \rightarrow 3 \rightarrow 4 \rightarrow 2$ with a length of 3,
$A_5 : 1 \rightarrow 5 \rightarrow 3 \rightarrow 4 \rightarrow 2$ with a length of 4,
$A_6 : 1 \rightarrow 3 \rightarrow 4 \rightarrow 5 \rightarrow 2$ with a length of 4,
$A_7 : 1 \rightarrow 3 \rightarrow 5 \rightarrow 4 \rightarrow 2$ with a length of 4.

However, the path $A : 1 \rightarrow 5 \rightarrow 4 \rightarrow 3 \rightarrow 5 \rightarrow 2$ is not a minimal path. The reasons are presented as follows:

(1) The node 5 is included twice;
(2) its length equals to 5 which is the number of nodes in the graph;
(3) it is still a path after removing the edges $5 \rightarrow 4 \rightarrow 3 \rightarrow 5$.

Assuming that C is an edge set of a graph G if all edges in C are removed from G and no path exists between the source and the sink, C is called a cut from the source node to the sink node. Similarly, C is called a minimal cut when it is not a cut if any edge is removed from it. The number of edges in a minimal cut is called its order. For example, six minimal cuts exist in the graph shown in Fig. 6.2. They are:

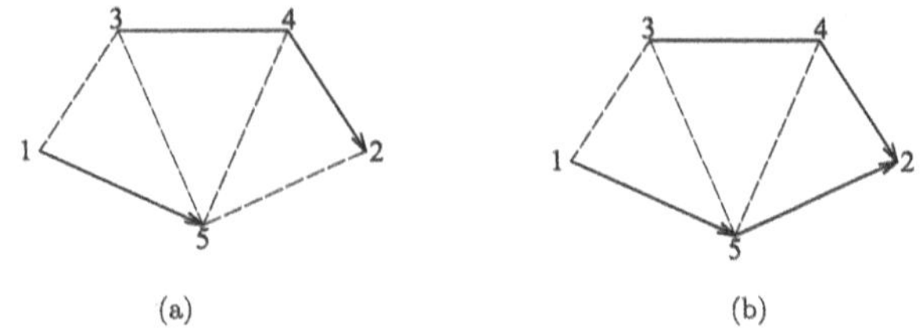

Fig. 6.3 Minimal cut

$C_1 : \{1 \rightarrow 3, 1 \rightarrow 5\}$ with an order of 2,
$C_2 : \{4 \rightarrow 2, 5 \rightarrow 2\}$ with an order of 2,
$C_3 : \{3 \rightarrow 4, 3 \rightarrow 5, 1 \rightarrow 5\}$ with an order of 3,
$C_4 : \{3 \rightarrow 4, 4 \rightarrow 5, 5 \rightarrow 2\}$ with an order of 3,
$C_5 : \{1 \rightarrow 5, 3 \rightarrow 5, 4 \rightarrow 5, 4 \rightarrow 2\}$ with an order of 4,
$C_6 : \{1 \rightarrow 3, 3 \rightarrow 5, 4 \rightarrow 5, 5 \rightarrow 2\}$ with an order of 4.

Take C_6 as an example to illustrate the characteristics of cuts and minimal cuts. According to the definition of cuts, when all edges in C_6 are removed, the original graph in Fig. 6.2 becomes the graph in Fig. 6.3a. Apparently, no path exists between the source 1 and the sink 2 in the graph. Thus, C_6 is a cut. However, when a new edge set is formed after removing edge $5 \rightarrow 2$ from C_6, i.e. edge $5 \rightarrow 2$ is added to the graph in Fig. 6.3a to form a new graph in Fig. 6.3b, a minimal path A_1 exists and, therefore, the new edge set is not a cut. Similarly, an edge set formed by removing any edge from C_6 is not a cut. Thus, C_6 is a minimal cut.

Usually, all minimal paths between the nodes v_1 and v_2 are called a minimal path set between those nodes and can be represented by $\{A_1, A_2, \ldots, A_m\}$. Similarly, all minimal cuts between two nodes are called a minimal cut set and can be represented by $\{C_1, C_2, \ldots, C_k\}$. A minimal cut set can be transformed into a minimal path set and vice versa.

6.2.2 *Structural Function of Network Systems*

For a network system, each element may be either in the operative state or failed state. Correspondingly, the network also owns two states, operative state or failed state. Using the Boolean algebra operators (see Appendix A), the operative state and the failed state can be represented by 1 and 0, respectively.

Any network consists of a lot of elements. Thus, when analyzing network reliability, the logical relationships among these elements must be determined. The expression of the relationship is called a structural function. When the network S consists of n elements, its structural function can be expressed as follows:

$$S = \Phi(x_1, x_2, \ldots, x_n) = \Phi(X) \tag{6.2}$$

where $\Phi(\cdot)$ is a general function and x_i is a Boolean variant:

$$x_i = \begin{cases} 1 & \text{if unit } i \text{ opeartes} \\ 0 & \text{if unit } i \text{ fails} \end{cases} \tag{6.3}$$

Apparently,

$$\bar{x}_i = \begin{cases} 0 & \text{if unit } i \text{ opeartes} \\ 1 & \text{if unit } i \text{ fails} \end{cases} \tag{6.4}$$

When the element state is expressed by Boolean variants, the state of the system S can also be expressed by Boolean variant as follows:

$$S = \Phi(G) = \begin{cases} 1 & \text{if network operates} \\ 0 & \text{if network fails} \end{cases} \tag{6.5}$$

For example, a series system with n elements is shown in Fig. 6.4. The system operates if all elements are in operative states. Thus, its structural function is:

$$\Phi(X) = x_1 x_2 x_3 \ldots x_n = \prod_{i=1}^{n} x_i \tag{6.6}$$

Similarly, a parallel system with n elements is shown in Fig. 6.5. The system fails if all elements are in failed states. Thus, its structural function is:

$$\Phi(X) = 1 - \bar{x}_1 \bar{x}_2 \bar{x}_3 \ldots x_n = 1 - \prod_{i=1}^{n} \bar{x}_i \tag{6.7}$$

Originally, the structural function of a system is expressed by event expression. In this way, an element or a set of elements of a system are defined as events. Then

Fig. 6.4 A series system with n elements

Fig. 6.5 A parallel system

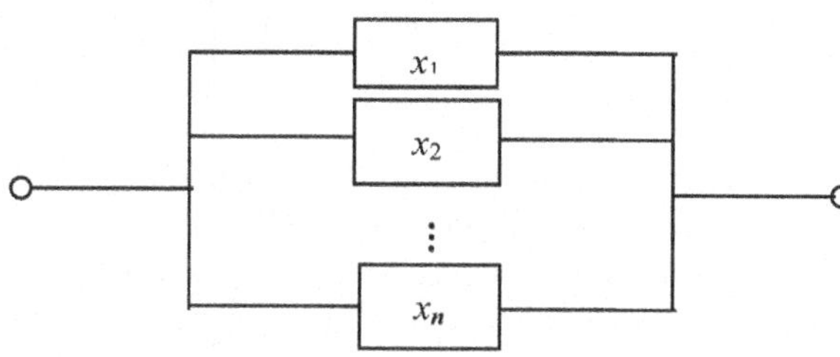

for a series network as shown in Fig. 6.4, the operation state of the system can be expressed as follows:

$$\Phi(X=1)=\bigcap_{i=1}^{n}(x_i=1) \tag{6.8}$$

while that the system fails means:

$$\Phi(X=0)=\bigcup_{\mathrm{i}=1}^{\mathrm{n}}(x_i=0) \tag{6.9}$$

Obviously, for complex systems, these expressions may bring unnecessary troubles. Therefore, it is better to introduce Boolean operators. In fact, for a series system, the structural function can be expressed as follows:

$$\Phi(X)=\bigcap_{i=1}^{n}x_i \tag{6.10}$$

For a parallel system, the structural function is

$$\Phi(X)=\bigcup_{\mathrm{i}=1}^{\mathrm{n}}x_i \tag{6.11}$$

As a complex two-terminal network can be expressed as the minimal path set in parallel, and each minimal path is a series system of corresponding elements. Thus, the structural function of a general network can be expressed as follows:

$$\Phi(X)=\bigcup_{i=1}^{m}A_i=1-\prod_{i=1}^{m}\left(1-\prod_{j=1}^{j_i}x_{ij}\right)(x_{ij}\in A_i) \tag{6.12}$$

Meanwhile, the two-terminal network can also be expressed as the minimal cut set in series, and each minimal cut is a parallel system of corresponding elements. Thus, the structural function can also be expressed as follows:

$$\Phi(X)=\bigcap_{l=1}^{k}C_l=\prod_{l=1}^{k}\left(1-\prod_{j=1}^{j_l}x_{lj}\right)(x_{lj}\in C_l) \tag{6.13}$$

However, these expressions do not mean that a complex network can be equally expressed as common series and parallel systems. The reason lies in that in common series and parallel systems, an element will not appear in different paths or cuts simultaneously. In other words, the corresponding Boolean variant of an element will appear only once in such kind of structural function. However, for a complex network, an element will appear in different paths or cuts many times. Usually, an element appearing many times means that the importance of the element is noteworthy.

6.2.3 Reliability of Simple Network System

The above expression is for a deterministic system. For a stochastic system, whether the system operates or not should be described by probability. Let P represents probability, then we will have:

$$\mathrm{P}(S) = \mathrm{P}(\Phi(X)) \tag{6.14}$$

The probability that the element operates can be expressed as follows:

$$p_i = \mathrm{P}(x_i = 1) \tag{6.15}$$

It is worth to point out that the probability $\mathrm{P}(g_i(\theta) > 0)$ or $\mathrm{P}(g_i(\theta) \leq 0)$ can be derived by structural reliability analysis methods such as Sect. 4.4. In fact, by introducing the state function of a network element, there exists

$$(x_i = 1) = (g_i(\theta) > 0) \tag{6.16}$$

and

$$(x_i = 0) = (g_i(\theta) \leq 0) \tag{6.17}$$

where $\theta = (\xi_1, \ldots, \xi_n)$ is related to random variables.

Therefore, reliable probability of a network element is

$$\mathrm{P}(x_i = 1) = \mathrm{P}(g_i(\theta) > 0) \tag{6.18}$$

and failure probability of a network element is

$$\mathrm{P}(x_i = 0) = \mathrm{P}(g_i(\theta) \leq 0) \tag{6.19}$$

Because Boolean variant x_i takes only two values, 0 and 1, the mathematical expectation of x_i is

$$E(x_i) = 1 \times \mathrm{P}(x_i = 1) + 0 \times \mathrm{P}(x_i = 0) = \mathrm{P}(x_i = 1) = p_i \tag{6.20}$$

Correspondingly, for stochastic system, its reliability is

$$\mathrm{P}(S) = P(S = 1) = E(S) = E[\Phi(X)] \tag{6.21}$$

When the random variables are s-independent, Eq. (6.21) can be expressed as the product of a lot of integrations of single variable. Then, for a simple series system, we have

$$\mathrm{P} = E(\Phi(x)) = E\left(\prod_{i=1}^{n} x_i\right) = \prod_{i=1}^{n} E(x_i) = \prod_{i=1}^{n} p_i \tag{6.22}$$

Apparently, because $p_i \leq 1$, more elements mean lower system reliability for series system.

For a parallel system, there exists

$$\mathrm{P}(S) = E\left(1 - \prod_{i=1}^{n} (1 - x_i)\right) = 1 - \prod_{i=1}^{n} (1 - p_i) \tag{6.23}$$

That means, for parallel system, more elements mean higher system reliability.

The above two expressions mean that, for a s-independent simple system, the expression of structural function is similar to that of reliability, where the element variables are replaced by element reliabilities. According to such similarity, system reliability can be got easily. Unfortunately, such convenience does not exist in complex networks. That is because that, for complex networks expressed in Eq. (6.12), one element may appear in different paths or cuts. Therefore, different paths A_i or cuts C_i may be dependent on each other. As a result, the above rule cannot be applied to complex networks. This means that the expectation operator $E(\cdot)$ and Boolean operator $\cup(\cap)$ have no interchangeability. Therefore, although structural functions of complex networks can be expressed as Eqs. (6.12) and (6.13), it cannot be transformed into the corresponding reliability expressions. In fact, in order to remove the dependencies between different paths or cuts, the disjointing algorithm needs to be introduced. Before that, the method to get minimal paths should be described.

6.3 Minimal Path Algorithm

There are a number of classical methods to get minimal path of a network (Mei, Liao, & Sun, 1987). Here, only some popular methods are introduced.

6.3.1 Adjacent Matrix Algorithm

For a network with n nodes, the following adjacent matrix can be used to represent the connection of nodes.

$$C = [c_{ij}] \tag{6.24}$$

where

$$c_{ij} = \begin{cases} x & \text{An edge } x \text{ connects node } i \text{ and } j \\ 0 & \text{No edge connects node } i \text{ and } j \end{cases}$$

For example, the adjacent matrix for the bridge network in Fig. 6.1 is

$$C = \begin{bmatrix} 0 & 0 & a & c \\ 0 & 0 & 0 & 0 \\ 0 & b & 0 & e \\ 0 & d & e & 0 \end{bmatrix} \tag{6.25}$$

It can be found that the diagonal elements are all zeros. Meanwhile, the elements of the column corresponding to the source, the first column and those of the rows corresponding to the sink, the second row, are all zeros.

Herein, $c_{ij} \neq 0$ means a minimal path with length of 1 from the node i to the node j. Moreover, it can be found that the power of the adjacent matrix can give the paths between nodes.

The square of adjacent matrix can be defined as follows:

$$C^2 = [{c_{ij}}^2] \tag{6.26}$$

where

$${c_{ij}}^2 = \begin{cases} \sum_{k=1}^{n} c_{ik}c_{kj} & i \neq j \\ 0 & i = j \end{cases} \tag{6.27}$$

Herein, $c_{ij}^2 \neq 0$ means a minimal path with a length of 2 from the node i to the node j. Taking the bridge's network as the example again, the square of its adjacent matrix is

$$C^2 = \begin{bmatrix} 0 & ab + cd & ce & ae \\ 0 & 0 & 0 & 0 \\ 0 & ed & 0 & 0 \\ 0 & eb & 0 & 0 \end{bmatrix} \tag{6.28}$$

Apparently, the above matrix tells whether the minimal paths exist between nodes, how many minimal paths exist and which edges form the minimal paths.

Furthermore, the mth power of adjacent matrix can be defined as follows:

$$C^m = [c_{ij}{}^m] \tag{6.29}$$

where

$$c_{ij}{}^m = \begin{cases} \sum_{k=1}^{n} c_{ik} c_{kj}{}^{m-1} & i \neq j \\ 0 & i = j \end{cases} \tag{6.30}$$

Similarly, $c_{ij}^m \neq 0$ means a minimal path with length of m from the node i to the node j. For a network with n nodes, all minimal paths between nodes can be derived by calculating $C \sim C^{n-1}$. So, adjacent matrix algorithm can give all minimal paths between nodes. Usually, the following matrix is called network reachability matrix.

$$Q = C + C^2 + \cdots + C^{n-1} \tag{6.31}$$

It represents the reachability of any node-pair in the network.

The minimal paths between one source and one sink can be got from the above analysis. In fact, if the node i is the source and the node j is the sink, only the jth column of $C \sim C^{n-2}$ and $c_{ij}{}^{n-1}$ are needed to be calculated to give all minimal paths between the source and the sink.

Taking the bridge network in Fig. 6.1 as an example. In order to give all minimal paths between the nodes 1 and 2, only following elements of the matrixes should be calculated:

$$c_2{}^2 = \begin{Bmatrix} ab + cd \\ 0 \\ ed \\ eb \end{Bmatrix} \tag{6.32a}$$

$$c_{21}{}^3 = aed + ceb \tag{6.32b}$$

6.3.2 Depth First Search Algorithm

The adjacent matrix algorithm is suitable for small and medium-scale networks. For a large-scale networks, adjacent matrix algorithm requires large storage and takes a long time. Therefore, for medium and large-scale network, the depth first search (DFS) and a breadth first search (BFS) methods are usually introduced to give the minimal paths. The DFS method is suitable for searching all minimal paths while

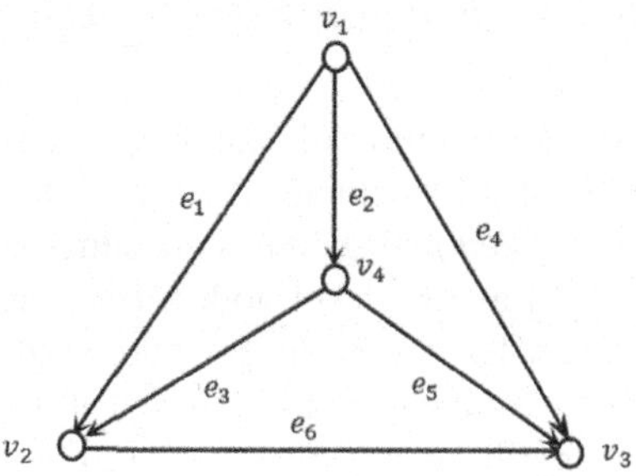

Fig. 6.6 An example network

the BFS method is suitable for searching the shortest minimal path. Herein, the DFS method is introduced firstly.

The basic idea of the DFS method is to start from the source v_0 and search for a neighbor node v_1. After the two nodes are marked, then an unmarked neighbor node v_2 of v_1 is searched and marked. The process is repeated until no marked neighbor node is available or it has arrived at the sink. If no marked neighbor node of v_i is available, return to its previous node v_{i-1} and search another unmarked node. If it has arrived at the sink, a minimal path is found. Apparently, the characteristic of the DFS method is searching in-depth firstly and return until no unmarked neighbor node is available.

Take the network in Fig. 6.6 as an example. The process of search a minimal path from the node v_1 to node v_4 is stated as following. From the node v_1, a neighbor node v_2 can be found and marked. Then an unmarked neighbor node v_3 of the node v_2 is found and v_3 is marked. However, as no unmarked neighbor of v_3 is available, it returns to its previous node v_2. Similarly, as no unmarked neighbor of the node v_2 is available, it returns to the node v_1, the previous node of v_2. Then, v_4, an unmarked neighbor node of v_1, is searched and a minimal path $(v_1, v_4) = e_2$ is found.

Lin provides a classic algorithm firstly to search all minimal paths of networks by applying the DFS method (Lin & Alderson, 1969). The process of the algorithm can be summarized as follows.

(1) Input the source node as a current node and set the states of the source as marked and all other nodes as unmarked;
(2) From the current node, select an unmarked neighbor node as the new current node and set its state as marked;
(3) Repeats (2) until no unmarked neighbor node of the current node is available or it has arrived at the sink. If no unmarked neighbor node of the current node is available, set the new current node as the previous node of the current node and return to (2); If it has arrived at the sink, go to (4);
(4) Record the minimal path.
(5) Judge whether all minimal paths have been found. If yes, end the algorithm. If no, set the new current node as the previous node of the current node and return to (2).

The above algorithm is suitable for directed, undirected and mixed networks.

6.3.3 *Breadth First Search Algorithm*

Another method of searching for a minimal path is the BFS algorithm. The basic idea of this method is to search minimal path level by level. Firstly, it sets source v_0 marked and searches all unmarked neighbor nodes of v_0. These neighbor nodes are set as first-level nodes, $\{v_{1i}\}$, and marked. Then, all unmarked neighbor nodes of first-level nodes in $\{v_{1i}\}$ are searched and are set as second-level nodes $\{v_{2i}\}$. The process repeats until it arrives at the sink.

It still takes the network in Fig. 6.6 as the example. Using the BFS method, the process of searching a minimal path from v_1 to v_4 is stated as following. From v_1, all neighbor nodes, i.e. the first-level nodes, v_2, v_3 and v_4 are found and marked. Then one shortest minimal path $(v_1, v_4) = e_2$ is found. If only one shortest minimal path is needed, the process ends. If all minimal paths are needed, the unmarked neighbor nodes, v_2 and v_3, are searched. As no unmarked neighbor node is available, the process ends and only one minimal path $(v_1, v_4) = e_2$ is found.

Apparently, the BFS method can search for the shortest minimal path between the source and the sink conveniently. The process of the BFS method can be stated as follows:

(1) Input the source as a current node and set its state as marked and the states of all other nodes as unmarked;
(2) From the current node, select all unmarked neighbor nodes as the next-level nodes and set their states as marked. If it has arrived at the sink, record the minimal path;
(3) Judge whether all same level nodes have been searched or not. If not, select the next same level node as the current node and return to (2). If yes, go to (4);
(4) Judge whether all nodes have been searched or not. If no, select a next-level node as the current node and return to (2). If yes, end the process.

Among the above process, the first minimal path must be the shortest minimal path.

6.4 Disjoint Minimal Path Algorithm

6.4.1 *Reliability Evaluation of Network System and Its Complexity*

Network system reliability is defined as the probability that the network system can satisfy a predetermined function at a given condition. If the predetermined function is the connectivity between a source node and a sink node, the network system reliability can be viewed as the node-pair connectivity reliability.

Let G be a network, v_1, v_2 are two given nodes, then, the connectivity reliability between v_1 and v_2 is called node-pair reliability which can be expressed by:

$$R = P(v_1 \text{can reach } v_2) \tag{6.33}$$

A connective system can be defined as follows:

$$S = (v_1 \text{can reach } v_2) \tag{6.34}$$

$$= (\text{at least one minimal path is available between } v_1 \text{ and } v_2)$$

That means there is at least one minimal path is between $v_1$andv_2, therefore:

$$\mathrm{S} = \bigcup_{i=1}^{m} A_i \tag{6.35}$$

Then the system reliability is

$$R = P(S) = P\left(\bigcup_{i=1}^{m} A_i\right) \tag{6.36}$$

Therefore, the connectivity reliability between two nodes can be theoretically determined if all minimal paths between the nodes are found. However, this apparently simple problem owns high complexity. Such complexity includes two aspects:

(1) Network structural complexity. The number of the minimal paths increases greatly with the growth of network size (number of nodes) and network complexity (node degree), which leads to the difficulty in storing all the minimal paths (insufficient space).
(2) Computation complexity. Because one identical element may appear in different minimal paths, there exist correlations among different A_i. In these cases, based on the probability theory, there exists

$$\begin{aligned} P\left(\bigcap_{i=1}^{m} A_i\right) &= \sum_{i=1}^{m} P(A_i) - \sum_{1 \le i < j = 2}^{m} P(A_i A_j) \\ &+ \sum_{1 \le i < j < k = 3}^{m} P(A_i A_j A_k) + \cdots + (-1)^{m-1} P\left(\bigcap_{i=1}^{m} A_i\right) \\ &= \sum_{i=1}^{m} (-1)^{j-1} \sum_{1 \le i_1 < i_2 \ldots < i_j = j}^{m} P\left(\bigcap_{l=1}^{j} A_l\right) \end{aligned} \tag{6.37}$$

This equation requires at least $2^{\mathrm{m}-1}$ summations even negelecting the product items, which obviously is a great number for a large network system. For example,

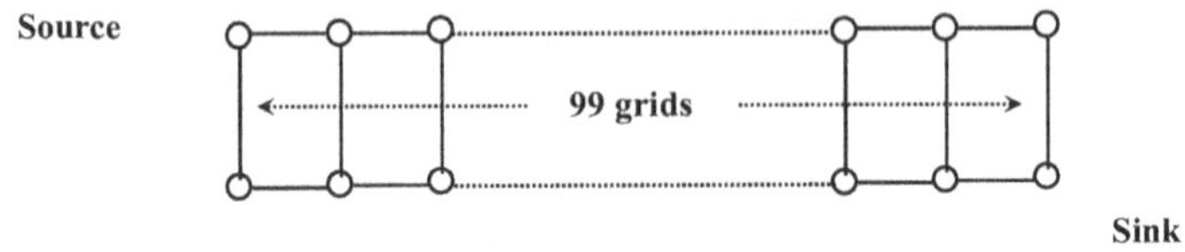

Fig. 6.7 A grid network

for a grid network with 200 nodes shown in Fig. 6.7, the number of the minimal paths is as high as 2^{99}. So, $2^{2^{99}} - 1$ summations are required based on Eq. (6.37). It is an impossible task even using a modern computer with a large memory and fast running speed.

6.4.2 *Disjoint Minimal Path Algorithm*

A method to decrease the computation complexity of the above problem is to adopt disjoint rules of sets (Liao, 1982a,1982b; Abraham, 1979).

Considering the union of two sets firstly

$$S = A_1 \bigcup A_2 \tag{6.38}$$

Figure 6.8 shows the sets A_1 and A_2. Apparently, $\bar{A}_1$ means the region inside the rectangle and outside the circle A_1. Thus, the shaded part is the intersection of $\bar{A}_1$ and A_2. Then, S can be expressed as the sum of disjointed set:

$$S = A_1 + \bar{A}_1 \ A_2 = B_1 + B_2 \tag{6.39}$$

The process of transferring Eqs. (6.38)–(6.39) is called the disjointed process. Using the disjointed technology, the general minimal paths of a network can be transformed into a set of disjointed minimal paths (DMP). For example, Eq. (6.33) can be transformed as follows:

Fig. 6.8 Disjoint set

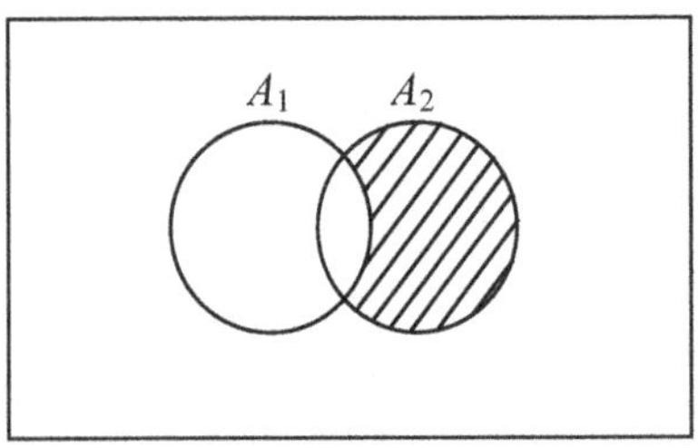

$$S=\bigcup_{i=1}^{m}A_i=A_1+\bar{A_1}\cdot A_2+\bar{A_1}\cdot\bar{A_2}\cdot A_3+\cdots+\bar{A_1}\cdot\bar{A_2}\cdots\bar{A}_{m-1}\cdot A_m$$

$$=A_1+\sum_{j=2}^{m}\left(\prod_{i=1}^{j-1}\bar{A_i}\right)A_j \tag{6.40}$$

Herein, the De. Morgan theorem in Appendix A can be used to expand $\left(\prod_{i=1}^{j-1}\bar{A_i}\right)A_j$. Generally, each $\left(\prod_{i=1}^{j-1}\bar{A_i}\right)A_j$ contains more than one term and each term is a disjoint minimal path. However, the direct expansion of $\left(\prod_{i=1}^{j-1}\bar{A_i}\right)A_j$ is complex. Following four theorems in Liao's literature (Liao, 1982a,1982b) can be used to speed up the process.

Theorem 6.1 When A_i and A_j do not contain the same element, $\bar{A_i}\ A_j$ can be expanded directly.

For example, if $A_i=abc$ and $A_j=defg$, then

$$\bar{A_i}A_j=\overline{abc}\cdot defg$$

$$=(\bar{a}+a\bar{b}+ab\bar{c})defg=\bar{a}defg+a\bar{b}defg+ab\bar{c}defg=B_1+B_2+B_3$$

Theorem 6.2 When A_i and A_j contain the same elements, $\bar{A_i}\ A_j$ can be expressed as the following equation:

$$\bar{A_i}\ A_j=\bar{A}_{i\leftarrow j}A_j \tag{6.41}$$

where the subscript $i\leftarrow j$ means the same elements are removed and $A_{i\leftarrow j}$ contains the elements of A_i but not the elements of A_j.

For example, if $A_i=abcde$ and $A_j=abdfg$, then $A_{i\leftarrow j}=ce$. Thus, we have

$$\bar{A_i}\ A_j=\overline{ce}\cdot abdfg=\left(\bar{c}+c\bar{e}\right)abdfg=\bar{c}abdfg+c\bar{e}abdfg \tag{6.42}$$

Theorem 6.3 When A_i and A_k,A_j and A_k contain the same elements,$\bar{A_i}\cdot\bar{A_j}\cdot A_k$ can be expressed as the following equation:

$$\bar{A_i}\bar{A_j}\cdot A_k=\bar{A}_{i\leftarrow k}\cdot\bar{A}_{j\leftarrow k}\cdot A_k \tag{6.43}$$

Obviously, the theorem is an extension of the Theorem 6.2 .

Theorem 6.4 When A_i $and A_k, A_j$ and A_k contain the same elements and $A_{i\leftarrow k} \subset A_{j\leftarrow k}, \overline{A_i} \cdot \overline{A_j} \cdot A_k$ can be expressed as the following equation:

$$\overline{A_i} \cdot \overline{A_j} \cdot A_k = \overline{A_{j\leftarrow k}} \cdot A_k \tag{6.44}$$

For example, if $A_i = abcde, A_j = abdg$ and $A_k = bdgh$, then $A_{i\leftarrow k} = ace$ and $A_{j\leftarrow k} = a$. As$A_{i\leftarrow k} \subset A_{j\leftarrow k}$, $\overline{A_i}\overline{A_j} \cdot A_k = \overline{a}\ bdgh$

It should be pointed out that the goal of disjointing minimal paths is not to remove the same element in different minimal paths but to guarantee any two disjoint minimal paths contain reverse elements. In order to understand this, let us see the example following theorem 6.1. After disjointing, three disjointed minimal paths, B_1, B_2 and B_3, are given. Apparently, any two disjointed minimal paths contain converse elements. For example, the element $\overline{a}$ in B_1 and the element a in B_2 are converse elements. As the result, the intersection of any disjoint minimal paths is equal to zero. After disjointing minimal paths, Eq. (6.40) becomes

$$S = \bigcup_{i=1}^{m} A_i = A_1 + \sum_{j=2}^{m} \left(\prod_{i=1}^{j-1} \overline{A_i} \right) A_j = \sum_{i=1}^{M} L_i \tag{6.45}$$

6.4.3 *Reliability Analysis Based on DMP Algorithm*

Because L_i is a disjoint set, the network connectivity reliability of system S can be given by

$$R = P(S) = \sum_{i=1}^{N} P(L_i) \tag{6.46}$$

where N is the number of all disjoint minimal paths (DMPs). Generally, the number of disjoint minimal paths is more than that of minimal paths, i.e. $N > m$.

Therefore, the steps to give network connectivity reliability using the disjoint minimal path method include.

(1) Search all the minimal paths between source and link nodes by Lin method;
(2) Transform the minimal path sets $(A_1, A_2, \ldots, A_m)$ to the disjoint minimal path set $(L_1, L_2, \ldots, L_n)$ using the Theorems 6.1–6.4 and the Boolean algebra operators;
(3) Calculate $P(L_i)$ based on the element reliability;
(4) Calculate the network connectivity reliability using Eq. (6.46).

It is noted that, in the system reliability analysis, the element reliability is a basic requirement. In other words, the structural reliability which is described in Chaps. 4 and 5 is a foundation of the system reliability analysis.

6.5 Recursive Decomposition Algorithm

Using the disjoint minimal path method, the computational complexity caused by the correlations among different minimal paths is essentially solved by reducing the number of summation items in Eq. (6.37) from 2^{m-1} to N. Although N may be much larger than m, the total computation cost is still reduced obviously. However, the disjoint minimal path method cannot solve the complexity problem of network structure. Actually, in the above method, all the minimal paths between two nodes must be found and stored before disjointing, which makes it difficult to store all minimal paths for larger scale networks. For example, for the network with 60 nodes shown in Fig. 6.7, the number of minimal paths between source and sink node is as high as 2^{29}. Moreover, as $N > m$, the storage of all the minimal paths is nearly impossible. Actually, the computation cost in Eq. (6.46) will increase rapidly if m increases rapidly. In fact, for large-scale networks, two challenging problems of analyzing connectivity reliability of a network are still computational and space complexities even using the DMP algorithm. A recursive decomposition algorithm (RDA), proposed by Li and He (2002), provides an effective tool to calculate the connectivity reliability of large-scale networks.

6.5.1 Related Theorems

Before introducing the recursive decomposition algorithm, two related theorems are introduced.

Theorem 6.5 *Let G represent a connected network and a be an element of G. When element a fails, i.e. a = 0, it becomes network G_f. When element a operates, i.e. a = 1, it becomes network G_r. Then, it can give.*

$$\Phi(G) = \bar{a}\ \Phi(G_f) + a\Phi(G_r) \tag{6.47}$$

Generally, this theorem is called network decomposition theorem. Using the theorem, it can be proved that there exists following theorem.

Theorem 6.6 *Let Grepresent a connected network and $a_f$be an element of G. When element $a_f$fails, it can be deleted and the network Gbecomes network G_f. Then, there exists:*

$$\bar{a_f}\ \Phi(G) = \bar{a_f}\ \Phi(G_f) \tag{6.48}$$

6.5.2 RDA for Edge-Weighted Network

For the edge-weighted networks, the elements in the Theorems 6.5 –6.6 are edges. For an edge-weighted network, its structural function can be defined as follows:

$$\Phi(G) = \bigcup_{k=1}^{m} A_k \tag{6.49}$$

Define the shortest minimal path from the source node to one of the sink node of the network as $A_1 = a_{11}a_{12}\ldots a_{1m_1}$, where a_{1i} is the ith edge of the path and m_1 is the number of edges in A_1. A_1 can be searched by the BFS algorithm. According to the Boolean laws, the structural function $\Phi(G)$ in Eq. (6.49) can be rewritten as follows:

$$\Phi(G) = A_1 \bigcup (\bigcup_{k=1}^{m} A_k) = A_1 \cup \Phi(G) \tag{6.50a}$$

According to disjointing formula, i.e., Eq. (6.39), the above equation can be rewritten as follows:

$$\Phi(G) = A_1 + \bar{A}_1\Phi(G) \tag{6.50b}$$

Using De Morgan's Law, $\bar{A}_1$ can be expanded as follows:

$$\bar{A}_1 = \bar{a}_{11} + a_{11}\bar{a}_{12} + \cdots + a_{11}a_{12}\cdots a_{1i-1}\bar{a}_{1i} + \cdots + a_{11}a_{12}\cdots \bar{a}_{1m_1} \tag{6.51}$$

Substituting Eq. (6.51) into Eq. (6.50b) and using the theorems 6.6, structural function $\Phi(G)$ becomes

$$\begin{aligned}\Phi(G) = A_1 + \bar{a}_{11}\Phi(G_{11}) + a_{11}\bar{a}_{12}\Phi(G_{12}) + \cdots \\ + a_{11}a_{12}\cdots a_{1i-1}\bar{a}_{1i}\Phi(G_{1i}) + a_{11}a_{12}\cdots \bar{a}_{1m_1}\Phi(G_{1m_1})\end{aligned} \tag{6.52}$$

where G_{1i} represents a subgraph after removing edge a_{1i} from G.

The above subgraphs $G_{1i}(i = 1, 2, \ldots m)$ can be classified into connected subgraphs and disconnected subgraphs. If taking the number of connected subgraphs as m_{1c}, the number of disconnected subgraphs is $m_{1u} = m_1 - m_{1c}$. Obviously, the coefficient in front of disconnected subgraph is a disjoint minimal cut of G. Let $C_{1j} = a_{11}a_{12}\ldots\bar{a}_{1j}$, then the complementary structural function can be written as follows:

$$\Phi' = 1 - \Phi(G) = \sum_{j=1}^{m_{1u}} C_{1j} + Q_1 \tag{6.53}$$

where Q_1 is the remainder term, denoting that $C_{1j}(j = 1, 2, \ldots, m_{1u})$ does not include all the disjoint minimal cut set of the network.

Because the contribution of the disconnected subgraph for network connectivity reliability is zero, all the disconnected subgraph can be deleted from the structural function. Then $\Phi(G)$ can be rewritten as follows:

$$\Phi(G) = A_1 + \sum_{i=1}^{m_{1c}} C_{1i}\Phi(G_{1i}) \tag{6.54}$$

where $C_{1i} = a_{11}a_{12}\ldots a_{1i-1}\bar{a}_{1i}$ and G_{1i} denotes the first order connected subgraph.

Decompose the above connected subgraphs step by step until no connected subgraph exists, then all disjointed minimal paths will be enumerated. Finally, $\Phi(G)$ can be expressed as follows:

$$\Phi(G) = A_1 + \sum_{i=2}^{m_{1c}} B_i A_{2i} + \sum_{i=m_{1c}+1}^{m_{2c}} B_i A_{3i} + \cdots + \sum_{i=m_{(n-1)c}+1}^{m_{nc}} B_i A_{(n+1)i} \tag{6.55}$$

where B_i is a coefficient, A_{ki} is a shortest minimal path of the ith connected subgraph and $m_{nc} + 1$ is the number of all connected graphs.

Let $L_1 = A_1$, $L_{n+1} = B_i A_{ki}(k = 2, \ldots, n + 1; i = 2, 3, \ldots, m_{nc})$, then Eq. (6.55) can be transformed into

$$\Phi(G) = \sum_{i=1}^{M} L_i \tag{6.56}$$

where L_i is a disjoint minimal path and $M = m_{np} + 1$.

At the same time, all disjoint minimal cuts are also enumerated, that is

$$\Phi'(G) = \sum_{j=1}^{K} C_j \tag{6.57}$$

where C_j is a disjoint minimal cut and K is the number of all disjoint minimal cuts.

In Fig. 6.9, a simple bridge network is used to illuminate the process of the algorithm.

After enumerating all disjoint minimal paths and disjoint minimal cuts, the network reliability can be given by

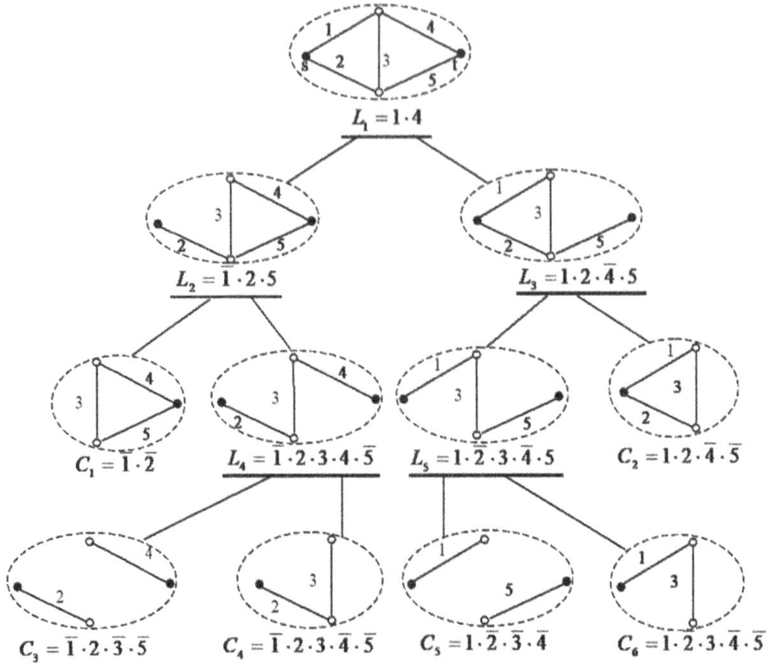

Fig. 6.9 Recursive decomposition algorithm for a bridge network

$$P_s = P[\Phi(G)] = \sum_{i=1}^{M} P(L_i) \tag{6.58}$$

and the network failure probability is

$$P_f = P\big[\Phi'(G)\big] = \sum_{j=1}^{K} P(C_j) \tag{6.59}$$

It should be noted that each disjoint minimal path or disjoint minimal cut is consisted of a series of edges whose reliability $\mathrm{P}(a_i)$ or failure probability $\mathrm{P}(\bar{a}_i)$ can be evaluated by the method described in Chap. 4. That means the structural reliability analysis is the foundation of network connectivity reliability evaluation.

For a large network, it might be impossible to enumerate all disjoint events. In these cases, the upper and lower boundaries of the network reliability can be calculated according to the probabilistic inequality:

$$\sum_{i=1}^{M_s} P(L_i) \le P_s \le 1 - \sum_{j=1}^{K_f} P(C_j) \tag{6.60}$$

where M_s and K_f are the numbers of the decomposed disjoint minimal paths and disjoint minimal cuts, respectively. Obviously, $M_s \leq M$ and $K_f \leq K$.

When the difference between the upper and the lower boundaries is less than a prescribed threshold, the network reliability can be given as follows:

$$P_s = \frac{1}{2}\left(1 + \sum_{i=1}^{M_s} P(L_i) - \sum_{j=1}^{K_f} P(C_j)\right) \tag{6.61}$$

Comparing with the disjoint minimal path method described in the above section, the RDA finishes the disjoint operation during the process of decomposition and therefore greatly simplifies the searching process of a disjoint path. Meanwhile, because every search can give a disjoint minimal path or a disjoint minimal cut, its probability can be added and therefore the disjoint minimal paths or the disjoint minimal cuts do not need to be stored. These advantages decrease complexity greatly. Meanwhile, as disjoint minimal paths and disjoint minimal cuts are derived simultaneously, it can give approximate high-precision solution without obtaining all disjoint minimal paths.

When programming the algorithm, the steps can be illustrated as follows:

(1) Let $R = 0$, $F = 0$, $i = 0$, $j = 0$.
(2) Decompose network G_i according to Eqs. (6.51–6.54), and then go to step (4);
(3) Judge whether or not the decomposed networks are connected graph. If yes, a shortest minimal path A_i is generated, and then go to step (2). Otherwise, go to step (5);
(4) $R = R + P(L_i)$, remove L_i, $i = i + 1$. Go to step (3);
(5) $F = F + P(C_j)$, remove C_j, $j = j + 1$;
(6) Judge whether the error between the upper and lower bound of the reliability is smaller than a threshold. If yes, go to step (7). Otherwise, go to step (3);
(7) Output R and F, and end the computation.

Figure 6.10 is an edge-weighted network with 34 nodes and 128 edges. Node 20 is the source node and node 25 is the sink node. Assume that the edge reliabilities are 0.9, 0.88 and 0.85, respectively, and all edges are s-independent. Using the recursive decomposition algorithm, the sink reliability with different error thresholds can be obtained and are shown in Table 6.1. In the table, the decomposed disjoint events and computation time are also presented. The example indicates that, when the error threshold is the same, with the decrease of the edge reliabilities, the decomposed disjoint events increase and the computation time increase as well. Furthermore, when the error threshold is in a certain range (0.5% in this example), the decomposed events and computation time are approximately a linear relationship with the error threshold. These conclusions are also true in general large-scale network analysis (Li & He, 2002).

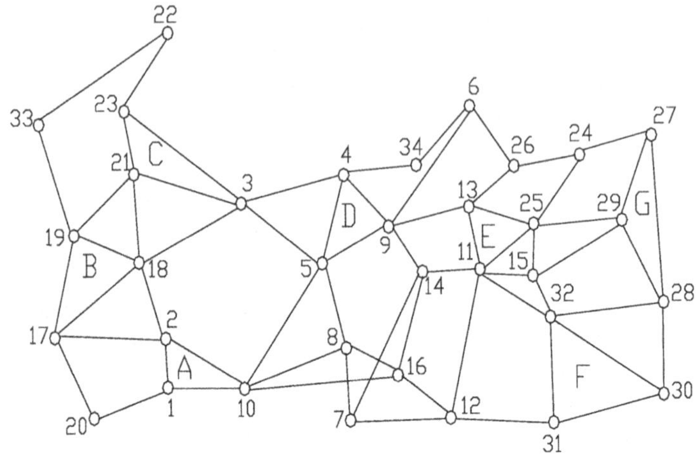

Fig. 6.10 The layout of the network

6.5.3 RDA for Node-Weighted Network

In lifeline engineering systems, there are a number of node-weighted networks. For example, when the failure of transmission lines is not considered, the electric power networks are node-weighted networks in which each node may be a transmission tower or an electric-voltage transfer substation.[1] For node-weighted networks, the basic system element is a structure or a structural system. The structural reliability analysis can use the physical synthesis method introduced in Chap. 5. Then the connectivity reliability of the node-weighted network can be analyzed by the recursive decomposition algorithm. However, considering the difference between edge-weighted network and node-weighted network, when employing the recursive decomposition algorithm, the generations of minimal paths and subgraphs should be changed as follows:

(1) The searched minimal path is the node minimal path, which do not consist of the source and the sink;
(2) When decomposing the networks, all the incident edges of the nodes corresponding to the Boolean variant $\overline{a_{1i}}$ should be removed from the original network.
(3) In the process of producing the decomposed disjoint minimal path L_i step by step, the approximate reliability of the node-weighted network can be given by:

$$R_S \approx \sum_{i=1}^{m_s} P(L_i) \tag{6.62}$$

[1]For the seismic reliability analysis of electric-voltage transfer substation, see Appendix B of the book.

Table. 6.1 Results of the edge-weighted network using RDA

Error threshold	Decomposed events			Computation time (s)			Reliability		
	Case I	Case II	Case III	Case I	Case II	Case III	Case I	Case II	Case III
0.0250	99	166	703	0.10	0.10	0.35	0.96882	0.96460	0.95672
0.0100	402	747	3752	0.15	0.30	1.0	0.98135	0.9766	0.96652
0.0040	1233	2921	18,263	0.45	1.0	4.0	0.98999	0.98094	0.97037
0.0020	2657	9015	42,862	1.0	2.3	9.0	0.98755	0.98221	0.97153
0.0010	7074	23,265	82,533	2.0	5.0	19.0	0.98816	0.98277	0.97181
0.0005	18,122	93,154	487,792	4.0	9.0	186.0	0.98853	0.98306	0.97180

Correspondingly, during the decomposition process, the disjoint minimal cut $C_j$are derived gradually, and the approximate failure probability is

$$F_S \approx \sum_{j=1}^{k_s} P(C_j) \tag{6.63}$$

For each L_i and C_i, the element reliability in the subset can be derived from the structural reliability analysis which is described in Chap. 5. For small and medium networks, all disjoint minimal paths and cuts can be derived, and therefore Eqs. (6.62) and (6.63) give the precise results of reliability and failure probability. For large-scale networks, considering the reliabilities of the source and the sink, there is the following approximate evaluation:

$$P(a_I)P(a_L)\sum_{i=1}^{m_s} P(L_i) \leq R \leq P(a_I)P(a_L)\left[1 - \sum_{j=1}^{k_s} P(C_j)\right] \tag{6.64}$$

where $P(a_I)$ is the reliability of the source and $P(a_L)$ is the reliability of the sink.

Then, when the error between lower and upper bound of Eq. (6.64) is smaller than a threshold, the approximate reliability can be given using the point estimation method, i.e. the mean of the lower and upper bound.

The network shown in Fig. 6.10 is taken as an example. But now it is a node-weighted network, and all node reliabilities are 0.9, 0.7 and 0.5, respectively. All nodes are s-independent. Using the RDA, the connectivity reliability of the node-weighted network is calculated and the results are shown in Table 6.2. From Table 6.2 , it is found that the computation efficiency of the node-weighted network is higher than the edge-weighted network.

Figure 6.11 is an actual power electricity network with 15 power factories, 37 power substations and 160 electricity lines. This network can be viewed as a node-weighted network. Nodes 2–16 are power factories and can be viewed as sources. For each node, it is actually a transform substation. The reliability analysis method of these systems can be found in Appendix B and analytical results are shown in Table 6.3. In the analysis, the seismic wave propagation and attenuation are considered and the results are shown in Table 6.3.

Using the RDA, the connectivity reliability of the network is calculated. The results are shown in Table 6.4. Meanwhile, the MCS method is employed and the number of simulations is 100,000. The results are also shown in Table 6.4 as well (Li, He, & Li, 2002). Obviously, the comparison validates the precision of RDA.

Table. 6.2 Results of the node-weighted network using recursive decomposition algorithm

Error threshold	Decomposed events			Computation time (s)			Reliability		
	Case I	Case II	Case III	Case I	Case II	Case III	Case I	Case II	Case III
0.025	23	906	1205	<1	1	1	0.77352	0.32557	0.06600
0.01	73	2469	2928	<1	1	1	0.78219	0.31602	0.05436
0.004	205	4868	6634	1	1	2	0.78701	0.31165	0.04630
0.002	345	6600	8688	1	2	2	0.78759	0.31006	0.04436
0.001	695	9222	11,553	1	2	2	0.78768	0.30924	0.04340
0.0004	1140	11,701	14,214	1	2	2	0.78764	0.30874	0.04281
0.005%	3127	17,691	19,713	1	3	3	0.78752	0.30844	0.04247
0.002%	5486	19,820	21,218	1	3	3	0.78751	0.30841	0.04244
0.0005%	7498	23,817	23,109	2	3	3	0.78750	0.30839	0.04243
0.0001%	11,111	24,424	24,014	2	3	3	0.78750	0.30839	0.04243

Table. 6.3 Node Reliability of the power electricity network

Node	Reliability	Node	Reliability	Node	Reliability	Node	Reliability
2	0.9996	15	0.9999	28	0.9999	41	0.9994
3	0.3409	16	0.9999	29	0.9999	42	0.9994
4	0.3500	17	0.9994	30	0.9994	43	0.9999
5	0.5878	18	0.8415	31	0.9999	44	0.9999
6	0.9996	19	0.9994	32	0.9999	45	0.9954
7	0.9996	20	0.9994	33	0.9900	46	0.9999
8	0.9999	21	0.6000	34	0.6461	47	0.9998
9	0.9999	22	0.6000	35	0.3154	48	0.9999
10	0.9999	23	0.6000	36	0.9904	49	0.9999
11	0.3409	24	0.7584	37	0.6591	50	0.9999
12	0.9996	25	0.7584	38	0.6914	51	0.9999
13	0.9996	26	0.9994	39	0.6914	52	0.9928
14	0.9999	27	0.9999	40	0.9954	53	0.9999

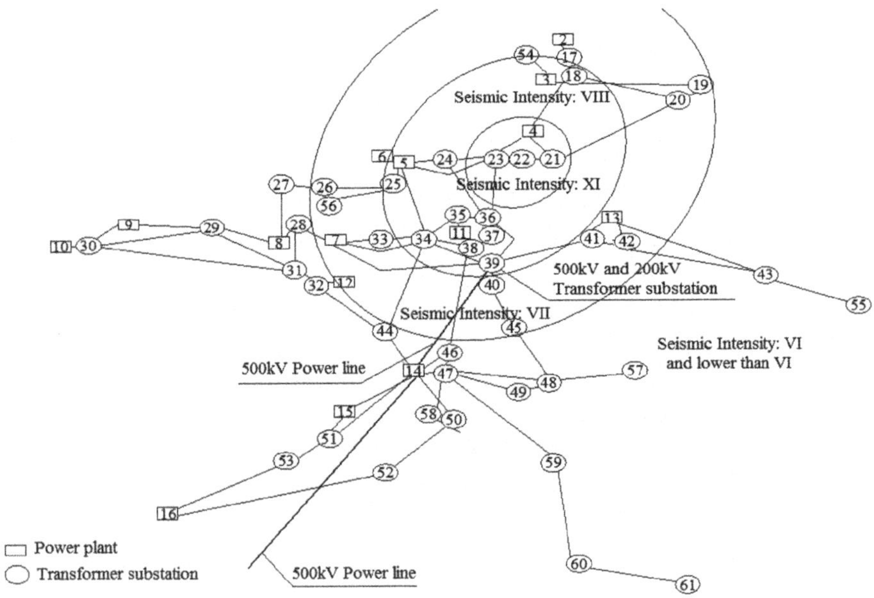

Fig. 6.11 A power electricity network

Table. 6.4 Node connectivity reliability of power electricity network

Node	RDA	MCS	Node	RDA	MCS
17	0.999222	0.9992	36	0.933726	0.9339
18	0.841213	0.8415	37	0.635966	0.6354
19	0.930118	0.9298	38	0.691385	0.6915
20	0.930451	0.9301	39	0.6914	0.6928
21	0.572815	0.573	40	0.993881	0.9939
22	0.48805	0.4871	41	0.999277	0.9992
23	0.585205	0.5863	42	0.999276	0.9993
24	0.730207	0.73	43	0.999776	0.9998
25	0.7584	0.7572	44	0.9999	0.9999
26	0.99938	0.9994	45	0.995293	0.9953
27	0.99989	0.9999	46	0.999886	0.9999
28	0.9999	0.9999	47	0.999786	0.9998
29	0.9999	0.9999	48	0.999823	0.9998
30	0	0	49	0.999823	0.9998
31	0.9999	0.9999	50	0.999899	0.9999
32	0.9999	0.9999	51	0.9999	0.9999
33	0.98986	0.99	52	0.9928	0.9929
34	0.6461	0.6453	53	0.9999	0.9999
35	0.309194	0.3083			

6.6 Cut-Based Recursive Decomposition Algorithm

When element reliabilities are low, for example, the element reliability is lower than 0.2, the computational efficiency of the RDA will decrease. This is because, for the specified error threshold, a huge number of disjoint minimal paths and cuts need to be searched for and decomposed. Therefore, for those networks with low edge reliabilities, a cut-based RDA may be a better choice (Liu & Li, 2012; Li, Qian, & Liu, 2007).

6.6.1 Minimal Cut Searching Algorithm

Merging node method (Lin, Kuo, & Yeh, 2003) is a better way to search for minimal cuts of a network. The basic idea of the method is to merge nodes into source set gradually and search all incident edges of the source to form minimal cuts.

The process of the method can be described as follows:

(1) Let the initial source set contain all sources. Remove the edge between the source and the redundant node and search all incident edges of the sources in the source set to form a minimal cut;
(2) Produce several networks that are all formed by merging one neighbor node of a source into the source set. Remove the edge between the source and the redundant node and search all incident edges of the sources in the source set to form a minimal cut. Judge whether this minimal cut is the same with previous minimal cuts or not. If not, add it to the minimal cut set. Otherwise, discard it;
(3) Repeat step (2) until the sink is in the source set.

Herein, the redundant node is the node which cannot arrive at the sink without crossing the node in the source set.

A network shown in Fig. 6.12 is used to illustrate the method. Herein, set *ss* is the source set.

Firstly, $ss = \{1\}$ corresponding to G_1, a minimal cut $\{e_1, e_2\}$ can be founded. Then two networks, G_2 and G_3, are formed by merging the nodes 2 and 3 into the source set, respectively. The source set of G_2 is $ss = \{1, 2\}$ and a minimal cut $\{e_2, e_3, e_4\}$ is found. Meanwhile, the source set of G_3 is $ss = \{1, 3\}$ and a minimal cut $\{e_1, e_5\}$ is found. Repeat the above steps until all the sink nodes are merged into source set. Finally, six minimal cuts are obtained. They are $\{e_1, e_2\}$, $\{e_2, e_3, e_4\}$, $\{e_1, e_5\}$, $\{e_3, e_4, e_5\}$, $\{e_1, e_4, e_6\}$, $\{e_3, e_6\}$.

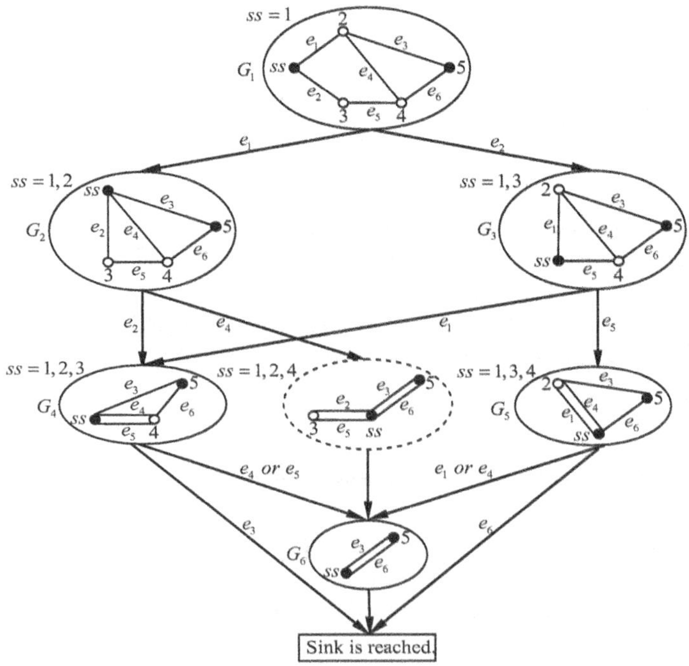

Fig. 6.12 An example network

The process of disjointing minimal cuts is similar to that of disjointing minimal paths. Actually, the four theorems for disjointing minimal path are also suitable for disjointing minimal cut. Let's take the Theorem 6.2 as an example here. Assuming $D_i = \bar{a}\bar{b}\bar{c}\bar{d}\bar{e}$ and $D_j = \bar{a}\bar{b}\bar{d}\bar{f}\bar{g}$, then $D_{i\to j} = \bar{c}\bar{e}$. Thus,

$$
\begin{aligned}
\bar{D}_i\, D_j &= \overline{\bar{a}\bar{b}\bar{c}\bar{d}\bar{e}} \cdot \bar{a}\bar{b}\bar{d}\bar{f}\bar{g} = (a + \bar{a}\, b + \bar{a}\bar{b}\, c + \bar{a}\bar{b}\bar{c}\bar{d}\, e)\, \bar{a}\bar{b}\bar{d}\bar{f}\bar{g} \\
&= (c + \bar{c}\, e)\, \bar{a}\bar{b}\bar{d}\bar{f}\bar{g} = \bar{D}_{i\to j} D_j
\end{aligned}
\tag{6.65}
$$

After getting all disjoint minimal cuts, Eq. (6.59) can be used to calculate the network failure probability. When the element failure is *s*-independent, the failure probability of a disjoint minimal cut can be calculated.

6.6.2 Cut-Based Recursive Decomposition Algorithm

According to the idea of the minimal cut, the complementary structural function of an edge-weighted network can be defined as follows:

$$
\Phi'(G) = \bigcup_{k=1}^{n} C_k \tag{6.66}
$$

Define an arbitrary minimal cut of a network as $D_1 = \bar{a}_{11}\bar{a}_{12}\ldots\bar{a}_{1m_1}$, where $\bar{a}_{1i}(i = 1, 2, \ldots, m_1)$ are the ith edges of the network, m_1 is the number of edges in D_1. According to the Boolean laws and disjointing formula (6.39), the complementary structural function $\Phi'(G)$ in Eq. (6.66) can be rewritten as follows:

$$
\Phi'(G) = D_1 \cup \left(\bigcup_{k=1}^{m} D_k \right) = D_1 \cup \Phi'(G) = D_1 + \bar{D}_1 \Phi'(G) \tag{6.67}
$$

Using the De Morgan's Law, $\bar{D}_1$ can be expanded as follows:

$$
\bar{D}_1 = a_{11} + \bar{a}_{11} a_{12} + \cdots + \bar{a}_{11}\bar{a}_{12} \cdots \bar{a}_{1i-1} a_{1i} + \bar{a}_{11}\bar{a}_{12} \cdots a_{1m_1} \tag{6.68}
$$

Substituting Eq. (6.68) into Eq. (6.67), $\Phi'(G)$ becomes

$$
\begin{aligned}
\Phi'(G) = D_1 &+ a_{11}\Phi'(G_{11}) + \bar{a}_{11} a_{12} \Phi'(G_{12}) + \cdots \\
&+ \bar{a}_{11}\bar{a}_{12} \cdots \bar{a}_{1i-1} a_{1i} \Phi'(G_{1i}) + \bar{a}_{11}\bar{a}_{12} \cdots a_{1m_1} \Phi'\left(G_{1m_1}\right)
\end{aligned}
\tag{6.69}
$$

where G_{1i} represents a subgraph after removing edge $a_{1j}(j = 1, 2, \ldots, i-1)$ from G and changing the edge a1i to an irremovable edge which means it cannot be in any cut set in the following steps.

The subgraphs $G_{1i}(i = 1, 2, \cdots, m)$ can be classified into two types. The first is the subgraphs in which a path exists between the source and the sink in which all edges are irremovable edges and the second is the subgraphs in which no path exists in which all edges are irremovable edges. Define the first as absorbed graph and the second as nonabsorbed graph. Apparently, no cut exists in the absorbed graph and the coefficients in front of an absorbed graph $\Phi'(G_{1j})$ is a minimal path as well. As the result, L_{1j} is a disjointed minimal path of G. Let $L_{1j} = \bar{a}_{11}\bar{a}_{12}\cdots\bar{a}_{lj-1}a_{lj}$, then the network structural function can be written as follows:

$$\Phi(G) = 1 - \Phi^{'}(G) = \sum_{j=1}^{m_{1a}} L_{1j} + Q_1 \tag{6.70}$$

where Q_1 is the remainder term, denoting that $L_{1j}(j = 1, 2, \ldots, m_{1a})$ cannot include all the disjoint minimal path set.

As no cut exists in absorbed graphs, they should be eliminated from the complementary structural function. Then the complementary structural function becomes

$$\Phi^{'}(G) = D_1 + \sum_{i=1}^{m_{1a}} L_{1i}\Phi^{'}(G_{1i}) \tag{6.71}$$

where $L_{1i} = \bar{a}_{11}\bar{a}_{12}\ldots\bar{a}_{1i-1}a_{1i}$ and G_{1i} denotes the first order non-absorbed subgraph.

Decomposing the non-absorbed subgraphs repeatedly until no non-absorbed graph exists, then all disjoint minimal cuts will be enumerated. Finally, the complementary structural function can be expressed as follows:

$$\Phi^{'}(G) = D_1 + \sum_{i=1}^{m_{1u}} E_i D_{i+1} + \sum_{i=m_{1u}+1}^{m_{2u}} E_i D_{i+1} + \cdots + \sum_{i=m_{(n-1)u}+1}^{m_{nu}} E_i D_{i+1} \tag{6.72}$$

where E_i is a coefficient, D_i is a minimal cut of the ith non-absorbed subgraph, $m_{nu} + 1$ is the number of all the non-absorbed subgraphs.

Let $C_1 = D_1$, $C_{i+1} = E_i D_{i+1}(1, 2, \ldots, m_{nu})$, then Eq. (6.72) becomes

$$\Phi^{'}(G) = \sum_{i=1}^{M} C_i \tag{6.73}$$

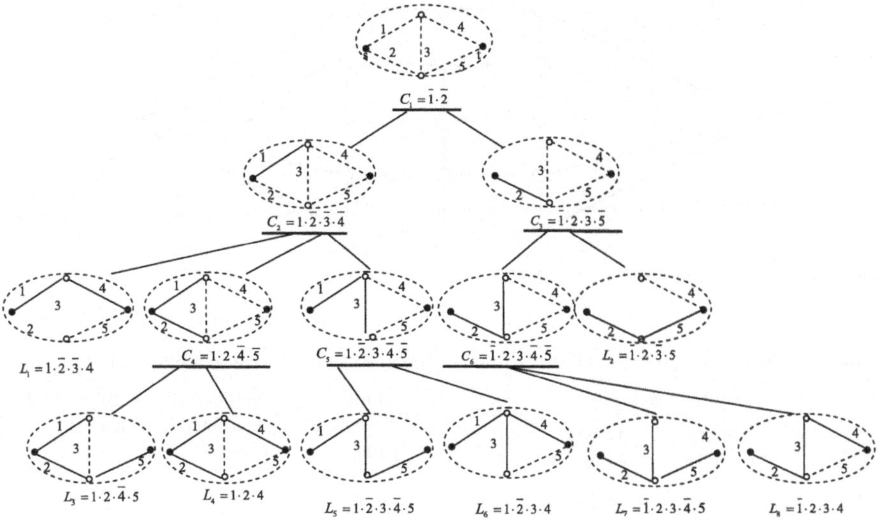

Fig. 6.13 A bridge network decomposed by the cut-based RDA

where C_i is a disjointed minimal cut and $M = m_{nu} + 1$.

Simultaneously, all disjoint minimal paths are enumerated as well, that is

$$\Phi(G) = \sum_{j=1}^{K} L_j \tag{6.74}$$

where L_j is a disjoint minimal path and K is the number of all disjoint minimal paths.

In Fig. 6.13, a bridge network is decomposed by the cut-based RDA.

After enumerating all disjoint minimal cuts and disjoint minimal paths, network failure probability should be:

$$P_f = P\left[\Phi^{'}(G)\right] = \sum_{i=1}^{M} P(C_i) \tag{6.75}$$

and the network reliability can be written as follows:

$$P_s = P[\Phi(G)] = \sum_{j=1}^{K} P(L_j) \tag{6.76}$$

Also, for large-scale networks, the probabilistic inequality can be used to give the upper and the lower boundaries:

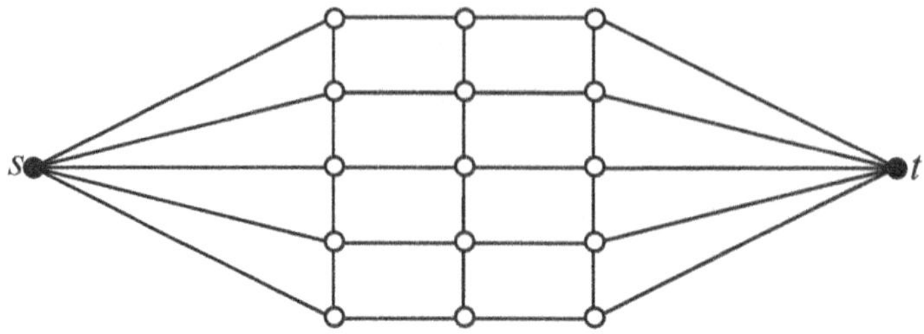

Fig. 6.14 A network with 17 nodes and 24 edges

Table. 6.5 Network failure probability using cut-based RDA with edge reliability 0.1

Threshold	Failure probability			Number of disjointed paths	Number of disjointed cuts	Computation time (s)
	Upper boundary	Lower boundary	Average			
0.001	0.99931	0.99834	0.99872	2322	4491	< 1
0.0005	0.99928	0.99878	0.99903	4797	7287	< 1
0.0002	0.99925	0.99905	0.99915	15,851	17,618	< 1
0.0001	0.99924	0.99914	0.99919	38,745	35,299	< 1

$$\sum_{j=1}^{K_s} P(L_j) \leq P_s \leq 1 - \sum_{i=1}^{M_f} P(C_i) \tag{6.77}$$

where K_s and M_f are the number of decomposed disjoint minimal paths and cuts, respectively.

When the difference between the upper and lower bounds is smaller than a prespecified value, the decomposition ends and the network reliability can be given as follows:

$$P_s = \frac{1}{2}\left(1 + \sum_{i=1}^{M_f} P(C_i) - \sum_{j=1}^{K_s} P(L_j)\right) \tag{6.77}$$

As an example, the network in Fig. 6.14 is calculated by the RDA, and the cut-based RDA respectively. Let all edges reliabilities be 0.1 and 0.3, respectively. For different error thresholds, the results using two algorithms are shown in Tables 6.5, 6.6,6.7 and 6.8. Obviously, for this example, when element reliability is low than 0.2, the cut-based RDA has a higher computational efficiency.

Apparently, the cut-based RDA can be extended for the node-weighted network.

Table. 6.6 Network failure probability using cut-based RDA with edge reliability 0.3

Threshold	Failure probability			Number of disjointed paths	Number of disjointed cuts	Computation time (s)
	Upper boundary	Lower boundary	Average			
0.001	0.8987	0.8977	0.8982	5,801,232	2,372,656	161
0.0005	0.8983	0.8978	0.8981	6,721,663	2,703,878	181
0.0002	0.8981	0.8979	0.8980	7,576,775	3,012,888	199
0.0001	0.8980	0.8980	0.8980	7,921,009	3,130,865	208

Table. 6.7 Network failure probability using RDA with edge reliability 0.1

Threshold	Failure probability			Number of disjointed paths	Number of disjointed cuts	Computation time (s)
	Upper boundary	Lower boundary	Average			
0.001	0.99922	0.99823	0.99872	855,883	435,340	44
0.0005	0.99922	0.99872	0.99897	1,122,044	730,524	75
0.0002	0.99922	0.99902	0.99912	1,373,420	1,131,735	96
0.0001	0.99922	0.99912	0.99917	1,504,480	1,437,828	109

Table. 6.8 Network failure probability using RDA with edge reliability 0.3

Threshold	Failure probability			Number of disjointed paths	Number of disjointed cuts	Computation time (s)
	Upper boundary	Lower boundary	Average			
0.001	0.8979	0.8969	0.8974	1,670,312	2,107,286	134
0.0005	0.8979	0.8974	0.8977	1,676,228	2,205,824	134
0.0002	0.8979	0.8977	0.8978	1,676,243	2,253,971	137
0.0001	0.8979	0.8978	0.8978	1,676,249	2,274,359	137

6.7 Reliability Analysis of Network with Dependent Failure

When the element failure is s-independent, Eq. (6.22) can be used to calculate the probability of a disjoint minimal path and Eq. (6.23) can do that for a disjoint minimal cut as it is a simple series system. However, when the element failures are dependent, the calculation of probability of a disjoint minimal path or a disjoint minimal cut will become much more complicated.

The dependences between different elements will cause two levels of dependence, i.e., the dependence between different minimal paths and dependence in one minimal path. The former comes from that one element (node or edge) may be included in different paths. The later comes from that the different elements in a minimal path may be failure dependent. After disjointing minimal paths, different disjoint minimal paths have incompatible elements. As a result, the product of two disjoint minimal paths must be zero, i.e. different disjoint minimal paths must be independent. Therefore, after disjointing, it only needs to consider the dependence of different elements in a disjoint minimal path.

The Ditleven boundary estimation method introduced in Chap. 4 can be used to calculate the probability of a disjoint minimal path or a disjoint minimal cut. However, this approach gives good results only when the correlation coefficient between the elements is smaller than 0.6. For general cases, an algorithm based on the condition reliability index can be used to deal with the dependent failure problem.

Assume the element reliability index is β and the element's performance function satisfies normal distribution. There exists

$$P(a_i = 1) = P(x_i < \beta) = \frac{1}{\sqrt{2\pi}} \int_{-\infty}^{\beta} \exp\left(-\frac{t^2}{2}\right) dt = \Phi(\beta) \tag{6.79a}$$

and

$$P\left(\bar{a}_i\right) = 1 - P(a_i) = 1 - \Phi(\beta) = \Phi(-\beta) \tag{6.79b}$$

Then the reliability of $L_1 = a_1 a_2 \cdots a_{K_i} \bar{a}_{K_i+1} \cdots \bar{a}_{K_i+N_l}$ is:

$$\begin{aligned} P(L_i = 1) &= P\left(\bigcap_{k=1}^{n} (x_k \le c_k)\right) \\ &= \int_{-\infty}^{c_n} \cdots \int_{-\infty}^{c_1} \frac{|\boldsymbol{R}|^{1/2}}{(2\pi)^{1/2}} \exp\left[-\frac{1}{2} \boldsymbol{X}^T \boldsymbol{R}^{-1} \boldsymbol{X}\right] dx_1 \cdots dx_n \end{aligned} \tag{6.80}$$

where $\boldsymbol{R}$ is the covariance matrix of random variables, $\mathbf{X} = (x_1, x_2, \ldots, x_n)^{\mathrm{T}}$ and

$$c_k = \begin{cases} \beta_k k \le K_i \\ -\beta_k k > K_i \end{cases} \tag{6.81}$$

According to condition probability, Eq. (6.80) can be rewritten as follows:

$$\begin{aligned} P(L_i = 1) = P(x_1 \le c_1) P(x_2 \le c_2 | x_1 \le c_1) \cdots \\ P(x_1 \le c_1) P\left(x_n \le c_n \middle| \bigcap_{k=1}^{n-1} (x_k \le c_k)\right) \end{aligned} \tag{6.82}$$

For the above condition probability, the following approximate equation exists (Singh, Vaithilingam, & Anne, 1996):

$$P\left(x_{n-r+1} \le c_{n-r+1} \middle| \bigcap_{k=1}^{n-r} (x_k \le c_k)(x_k \le c_k)\right) \approx \Phi(c_{n-r+1|n-r}) \tag{6.83}$$

where $c_{n-r+1|n-r}$ is the condition reliability index.

He and Li (2003) developed the above method and gave a general equation for above condition reliability index. That is:

$$c_{i|j-1} = \frac{c_{i|j-1} + \rho_{i,j|j-1} A_{j|j-1}}{\sqrt{1 - \rho_{i,j|j-1}^2 B_{j|j-1}}} \tag{6.84}$$

where

$$\rho_{i,j|j-1} = \frac{\rho_{i|j-2} + \rho_{i,j-1|j-2}\,\rho_{j,j-1|j-2}\,B_{j-1|j-2}}{\sqrt{1 - \rho_{i,j|j-1}^2 B_{j-1|j-2}}\sqrt{1 - \rho_{j,j-1|j-2}^2 B_{j-1|j-2}}} \tag{6.85}$$

$$A_{j|j-1} = \varphi(c_{j|j-1})\Phi^{-1}(c_{j|j-1}) \tag{6.86}$$

$$B_{j|j-1} = A_{j|j-1}(c_{j|j-1} + A_{j|j-1}) \tag{6.87}$$

where $\varphi(\cdot)$ is the probability density function of the standard normal distribution. Apparently, the above equation is a recursive equation from $j = 2$.

Thus, there exists

$$P(L_i = 1) \approx \prod_{k=1}^{n} \Phi(c_{k|k-1}) \tag{6.88}$$

For the disjoint minimal cuts, the same operator can also be adopted. Then the reliability of the network in which the elements are dependent can be given readily using the RDA.

6.8 Monte Carlo Simulation Method

The above algorithms, including the disjoint minimal path method, the disjoint minimal cut method, the RDA and the cut-based RDA are all probability analytical, i.e. they can give a precise solution of the network. On the other hand, there exists an approximate method, i.e., Monte Carlo simulation (MCS) method. The basic idea of the method is to give the operative frequency of the network by a certain number of simulations. With the increasing of the simulation times, the frequency is approaching to the exact network reliability.

Taking the edge-weighted network as an example, the process of the MCS method can be stated as follows:

(1) The reliabilities of network edges are given by using the structural reliability evaluation method in Chaps. 4 and 5;

(2) A series of random numbers between [0, 1] following with uniform distribution is produced and they are matched to edges of the network one by one, and give the random variable r_i;
(3) Compare random variable r_i with each edge reliability P_{r_i}. If $r_i \geq P_{r_i}$, the edge is in a safe state, otherwise, the edge fails;
(4) Use the BFS methods to judge whether the sinks are connected with the source or not according to each edge state;
(5) Repeat steps 2–4 and give the statistics of connected frequency of the network.

Generally, in order to give a stable connectivity reliability estimation, for a large-scale network, the simulations may need 10^4–10^6 times.

The most important advantage of the MCS method is simplicity. The connectivity reliability between the source and all sinks can be given simultaneously after a number of simulations. However, its disadvantages are also apparent:

(1) The convergent standard is hard to be determined. For a fixed standard, the number of simulations may vary greatly owing to different elements reliabilities.
(2) The MCS method is useful for the networks when the element failures are *s*-independent. However, when element failures are dependent, the MCS method will lose effectiveness.

References

Abraham, J. A. (1979). An Improved Algorithm for Network Reliability. *IEEE Transactions on Reliability, 1*(1), 58–61.

He, J., & Li, J. (2003). Approximate method for large engineering network system with correlation failure. *Chinese Journal of Computational Mechanics., 20*(3), 261–266 (in Chinese).

Liao, J. S. (1982a). A disjoint algorithm for network reliability (I). *Journal of Astronautics, 3,* 51–56 (in Chinese).

Liao, J. S. (1982b). A disjoint algorithm for network reliability (II). *Journal of Astronautics, 4,* 28–35 (in Chinese).

Li, J., & He, J. (2002). A recursive decomposition algorithm for network seismic reliability evaluation. *Earthquake Engineering and Structural Dynamics, 31*(8), 1525–1539.

Lin, M., & Alderson, G. E. (1969). Symbolic network functions by a single path-finding algorithm. In: *Proceedings of 7th Annual Allerton Conference on Circuit and System Theory* (pp. 196–205).

Lin, H. Y., Kuo, S. Y., & Yeh, F. M. (2003). Minimal cutset enumeration and network reliability evaluation by recursive merge and BDD. In *Proceedings of the Eighth IEEE Symposium on Computers and Communications* (pp. 1341–1346). ISCC 2003.

Liu, W., & Li, J. (2012). An improved cut-based recursive decomposition algorithm for reliability analysis of networks. *Earthquake Engineering & Engineering Vibration, 11*(1), 1–10.

Li, J., Qian, Y. K., & Liu, W. (2007). Minimal cut-based recursive decomposition algorithm for seismic reliability evaluation of lifeline networks. *Earthquake Engineering and Engineering Vibration, 6*(1), 21–28.

Mei, Q. Z., Liao, G. S., & Sun, H. Z. (1987). *Foundation of system reliability* (in Chinese), Science Press.

Singh, H., Vaithilingam, S., & Anne, R. K. (1996). Terminal reliability using binary decision diagrams. *Microelectronic Reliability, 36*(3), 363–365.

Chapter 7
Seismic Reliability Analysis of Engineering Network (II)—The Functional Reliability

7.1 Introduction

Unlike the connectivity reliability, the seismic functional reliability concerns about the probability of a engineering network satisfying the demand of customers. Many researchers regard the network connectivity reliability just as an approximation of the network functional reliability. However, this viewpoint might not be strictly correct. In fact, no matter in terms of the assessment object or the analysis method, the connectivity reliability and the functional reliability are two different concepts. Water supply system can be taken as an example to illustrate the difference. For such networks, the functional reliability is the probability that the water pressure satisfies customer's demand at one node, while the connectivity reliability represents the probability that at least a path exists between the source node and the customer node after an earthquake. Because of the potential leakage of water, the water pressure may decrease to zero before water flow arrives at the custom node although a path exists between the source and the custom node. Thus, the connectivity does not mean the function of the system satisfies the demand.

Owing to have to implement the functional analysis, the seismic functional reliability analysis is impossible without the support from the experts of specific field. Therefore, the researches on connectivity reliability analysis are much more than on the functional reliability analysis. Considering this background, only the functional reliability analysis of water supply networks, which is relatively mature in lifeline earthquake engineering, is introduced in this chapter.

7.2 Functional Analysis of Water Supply Network

The functional analysis of water supply networks is also known as the hydraulic analysis. From this analysis, the pipe flow and nodal head of a water supply network could be provided on the condition of knowing the topological structure of the network, pipe

J. Li and W. Liu, *Lifeline Engineering Systems*,
https://doi.org/10.1007/978-981-15-9101-3_7

parameters, nodal demand flow and source head, etc. Apparently, such an analysis provides the serviceability of a water supply network.

Three methods, namely the loop path method, the link path method and the node method, can be used to execute the hydraulic analysis of water supply networks (Yan and Zhao 1986). Generally, for large-scale networks, the node method is most suitable for giving nodal head, and pipe flow by solving a set of non-linear equations. These equations are derived by using the principle of mass conservation, the principle of energy conservation and the physical relationship between the flow and the pressure difference.

The connection relationships between pipes and nodes can be expressed by a connection matrix $\boldsymbol{A}$ which can be defined as follows:

$$\mathbf{A} = \left[a_{ij}\right] \tag{7.1}$$

where $\mathbf{A}$ is an $n \times m$ connectivity matrix and its element a_{ij} takes three possible values, namely, 0, 1 or -1. $a_{ij} = 0$ means there is no water in pipe j flows into or out of node i; $a_{ij} = -1$ means the water in pipe j flows in node i; and $a_{ij} = 1$ means the water in pipe j flows out of node i. Herein, m and n are the number of pipes and the nodes in water distribution networks, respectively.

As an example, the connection matrix for the water distribution network in Fig. 7.1 can be stated as follows:

$$\mathbf{A} = \begin{bmatrix} -1 & 1 & 1 & 1 & 0 & 0 & 0 & 0 & 0 & 0 \\ 0 & -1 & 0 & 0 & -1 & 0 & 1 & 0 & 0 & 0 \\ 0 & 0 & 0 & 0 & 0 & 0 & -1 & -1 & 0 & 1 \\ 1 & 0 & 0 & 0 & 0 & 0 & 0 & 0 & -1 & -1 \\ 0 & 0 & 0 & -1 & 0 & -1 & 0 & 1 & 1 & 0 \\ 0 & 0 & -1 & 0 & 1 & 1 & 0 & 0 & 0 & 0 \end{bmatrix}$$

The principle of mass conservation points out that, in a water supply network, the water flowing into a node is equal the water flowing out of the same node.

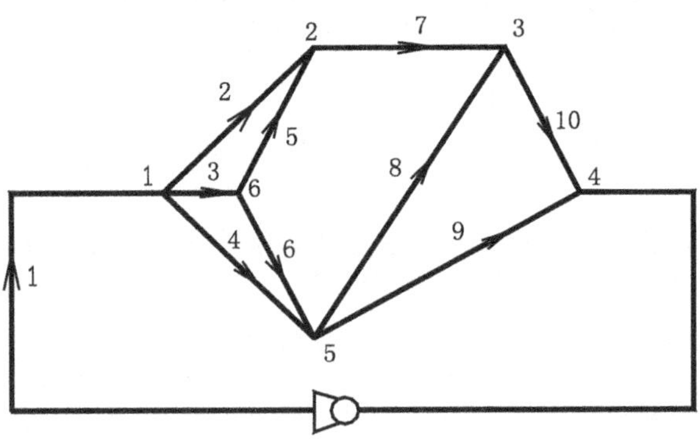

Fig. 7.1 A water distribution network

Accordingly, a nodal flow equilibrium equation can be given as follows:

$$\mathbf{AQ_P} = \mathbf{Q_N} \tag{7.2}$$

where $\mathbf{Q_P} = [q_{Pi}]\,(m^3/\text{sec})$ is an $m \times 1$ vector and each element represents the water flow transported in a pipe; $\mathbf{Q_N} = [q_{Ni}]\,(m^3/\text{sec})$ is an $n \times 1$ vector and each element represents the demand flow of a node. For example, the corresponding vectors of the water distribution network in Fig. 7.1 are $\mathbf{Q_P} = (q_1, q_2, \ominus, q_{10})^T$ and $\mathbf{Q_P} = (Q_1, Q_2, \ominus, Q_6)^T$, respectively.

Similarly, the principle of energy conservation requires that the head difference between two incident nodes of a pipe equals to the energy loss when water flows from one node to another node. For pipe *j* with two incident nodes, *i* and *k*, the energy loss can be written as follows:

$$\Delta E_j = H_i - H_k \tag{7.3}$$

where H_i and H_j are the heads of nodes *i* and *j*, respectively.

Using the above connection matrix, the energy loss of the network can be stated as follows:

$$\mathbf{E_P} = \mathbf{A^T H} \tag{7.4}$$

where $\mathbf{A}^T$ is the transpose matrix of $\mathbf{A}$, $\mathbf{H} = (h_1, h_2, ..., h_n)^T$ and $\mathbf{E_P} = (\Delta E_1, \Delta E_2, ..., \Delta E_m)$.

The relationship between the transmission flow of pipe and the head difference of its two incident nodes is a kind of physical relationship, which is usually obtained by experiments. There are many empirical equations. Usually, it can always be written as follows:

$$q_j = R_j(\Delta E_j)^\alpha \tag{7.5}$$

where R is a constant determined by pipe diameter, pipe length and pipe material, etc., α is a constant in the empirical equation. For example, when the Hazen-Williams equation is adopted, i.e.,

$$q = 0.278CD^{2.63}L^{-0.54}\Delta E^{0.54} \tag{7.6}$$

there are, $\alpha = 0.54$ and

$$R = 0.278CD^{2.63}L^{-0.54} \tag{7.7}$$

Table. 7.1 The value of C in Hazen-Williams equation (Yan and Zhao 1986)

Pipeline	C	Pipeline	C
Glass, plastic and steel pipelines	145–150	Welded steel pipeline-New	110
Cast iron pipeline-Excellent	140	Welded steel pipeline-Old	95
Cast iron pipeline-New	130	Rubber fire hose	110–140
Cast iron pipeline-Old	100	Concrete and asbestos cement pipelines	130–140
Cast iron pipeline-seriously corroded	90–100	PVC pipelines	150

where C is a coefficient and its values under different conditions are shown in Table 7.1, D (m) is the pipe diameter, L (m) is the pipe length, the unit of q is m^3/s and that of ΔE is m.

Substituting Eqs. (7.3) and (7.4) into Eq. (7.5) will give:

$$\mathbf{Q}_{\mathrm{P}} = \mathbf{R}\left(\mathbf{A}^{\mathrm{T}}\mathbf{H}\right)^{\alpha} \tag{7.8}$$

where $\mathbf{R} = (R_1, R_2, \ldots, R_m)^{\mathrm{T}}$。

Moreover, substituting Eqs. (7.8) in (7.2), there is

$$\mathbf{A}\mathbf{R}\left(\mathbf{A}^{\mathrm{T}}\mathbf{H}\right)^{\alpha} - \mathbf{Q}_{\mathrm{N}} = 0 \tag{7.9}$$

The above equation is an nodal hydraulic equation. Apparently, it is a set of non-linear equations of nodal head and can be solved by using numerical method.

7.3 Functional Analysis of Water Supply Network with Leakage

7.3.1 Hydraulic equation of water supply network with leakage

After earthquakes, pipes may be damaged and many breaks would occur on pipes. Therefore, the functional analysis of water supply networks after earthquakes is essentially a hydraulic analysis of network with leakage.

Taking the leakage into account, the nodal balance formulation of water supply networks, Eq. (7.2), can be modified to:

$$\mathbf{A}\mathbf{Q}_P - \mathbf{Q}_N - \mathbf{Q}_L = 0 \tag{7.10}$$

where $\mathbf{Q_L} = [\mathbf{q_{Lj}}](m^3/\sec)$ is an $\mathbf{n} \times 1$ vector and each element represents the leakage flow of the corresponding node.

Investigations during many past earthquakes indicated that the joints were the weakest components in water distribution networks. Thus, it is usually assumed that water leaks from the joints. There are different leakage models for joint leakage, a common formula is as follows:

$$q_L = \eta \mu A_L \sqrt{2gH_L} \tag{7.11}$$

where η is the leakage coefficient considering the impediment of soil and filling material on water flow and usually takes the value of 0.1–0.3; μ is the flow coefficient and takes the value of 0.6 for small orifice; H_L(m) is the head of a joint; g is the acceleration of gravity. Finally, $A_{L1}(\mathrm{m}^2)$ is the leakage area of the joint and can be written as follows (Chen and Li 2003):

$$A_{L1} = \begin{cases} 0 & S < R \\ \pi D(S - R_1) & R_1 \le S \le R_2 \\ \pi D(S - R_2) & R_2 \le S \le R_2 + D/4 \end{cases} \tag{7.12}$$

where D (m) is the diameter of the pipe; R_1 and R_2 are the joint leakage displacements defined in Table 4.5 for different types of pipes; S(m) is the joint displacement after earthquakes.

For a pipe with many joints, in order to improve the computational efficiency of flow analysis, it can be assumed that all the leakage flow of a pipe is concentrated on the incident nodes, and H_L is assumed to be equal to the mean head of two incident nodes. Then the leakage flow of a pipe can be written as follows:

$$q_L = \alpha \mu A_L \sqrt{g(H_1 + H_2)} \tag{7.13}$$

where H_1 and H_2 are the heads of two incident nodes and

$$A_L = N A_{L1} \tag{7.14}$$

where N is the joint number between two incident nodes.

7.3.2 Analysis method

Similar to the hydraulic analysis during daily operation, the nodal head of water supply networks with leakages can be obtained by solving the following equation:

$$\mathbf{AR}\left(\mathbf{A}^{\mathrm{T}}\mathbf{H}\right)^{\alpha} - \mathbf{Q}_{\mathrm{N}} - \mathbf{Q}_{\mathrm{L}} = 0 \tag{7.15}$$

The steps for the hydraulic analysis of water distribution networks with leakages can be stated as follows:

1. Give the connection matrix of networks and set the initial nodal head;
2. Substitute the head difference of two incident nodes of a pipe in Hazen-Williams equation to give pipe flow. Meanwhile, use the mean head of two incident nodes of a pipe to calculate the leakage flow.
3. Substitute pipe flow and node demand flow in the node flow equation to obtain node unbalanced flow. When the unbalanced flow is smaller than a prescribed small value ε, end the computation;
4. Differentiate the Hazen-Williams equation in terms of **H** to give $\frac{\partial Q_P}{\partial H}$ and differentiate the leakage model equation in terms of **H** to give $\frac{\partial \mathbf{Q}_{\mathrm{L}}}{\partial \mathbf{H}}$. Then calculate new nodal head using the following equation:

$$\mathbf{H}^{(\mathrm{n}+1)} = \mathbf{H}^{(\mathrm{n})} + \eta\left(\mathbf{A}\frac{\partial \mathbf{Q}_{\mathrm{P}}}{\partial \mathbf{H}} - \frac{\partial \mathbf{Q}_{\mathrm{P}}}{\partial \mathbf{H}}\right) \tag{7.16}$$

   where η takes the value of 0.2–0.6.
5. Repeat steps 2–4 until the prescribed accuracy is achieved.

7.4 Seismic Functional Reliability Analysis of Water Supply Network

In Eq. (7.12), the joint displacement S under earthquakes is actually a random variable. Obviously, S is related to seismic ground motion, pipe size, material elastic modulus and soil parameters, etc. Considering that the variances of pipe size and material elastic modulus are small, the randomness of S mainly comes from the randomness of seismic ground motions. Meanwhile, R_1 and R_2 in Eq. (7.13) are also random variables. Therefore, the probabilistic analysis is needed. The first-order reliability method (FORM) supplies a useful tool for this target.

The hydraulic Eq. (7.15) of the water supply networks with leakages can be rewritten as follows:

$$f(\mathbf{H}) = \mathbf{AR}\left(\mathbf{A}^{\mathrm{T}}\mathbf{H}\right)^{\alpha} - \mathbf{Q}_{N} - \mathbf{Q}_{L} = 0 \tag{7.17}$$

Let h be the inverse function of $\boldsymbol{f}(\boldsymbol{H})$. That is

$$H_i = r_i(\mathbf{A}, \mathbf{R}, \mathbf{Q}_N, \mathbf{Q}_L) \quad (i = 1, 2, \ldots, n) \tag{7.18}$$

Then the performance function of node j of the network can be expressed as follows:

$$Z_i = H_i - H_{i\min} \quad (i = 1, 2, \ldots, n) \tag{7.19}$$

where $H_{i\min}$ is the minimal demand head of node i which is customers' demands.

Accordingly, the seismic functional reliability of node i can be expressed as follows:

$$P_i = P(Z_i > 0) \tag{7.20}$$

In order to obtain such kind of reliability, in the first-order second-moment method, the performance function is expanded at the means of random variables and then approximate original faction as a linear function. This linear function is used to calculate the mean, the standard deviation and the reliability index of the performance function.

Let the basic random variable be A_{Lj}. Its mean and variance are, respectively:

$$\mu_j = E(A_{Lj}) = \overline{A_{Lj}} \tag{7.21}$$

$$\sigma_j^2 = E\left[(A_{Lj} - \mu_j)^2\right] \tag{7.22}$$

When the performance function is expanded at the means of random variables by Taylor's expansion and takes a linear approximation, the corresponding linear function is

$$Z_i = g_i(\mu_1, \mu_2, \ldots, \mu_n) - H_{i\min} + \sum_{j=1}^{n} a_j(A_{Lj} - \mu_j) \quad (i = 1, 2, \ldots, n) \tag{7.23}$$

where

$$a_j = \left(\frac{\partial g_i}{\partial Q_N} \frac{\partial Q_N}{\partial A_{Lj}} + \frac{\partial g_i}{\partial Q_L} \frac{\partial Q_L}{\partial A_{Lj}} \right)_{A_{Lj} = \mu_j} \tag{7.24}$$

Assuming all the random variables are independent, the mean and the standard deviation of the performance function will be, respectively:

$$\mu_{Z_i} = g_i(\mu_1, \mu_2, \ldots, \mu_n) - H_{i\min} \quad (i = 1, 2, \ldots, n) \tag{7.25}$$

$$\sigma_{Z_i} \approx \sqrt{\sum_{j=1}^{n} (a_j \sigma_j)^2} \quad (i = 1, 2, \ldots, n) \tag{7.26}$$

When all the random variables follow normal distributions, Z_i also follows normal distribution. Then the reliability of node i is

$$P_{si} = P(Z_i > 0) = \Phi(\beta_i) \quad (i = 1, 2, \ldots, n) \tag{7.27}$$

where $\Phi(\cdot)$ is the probability density function of the standard normal distribution, and

$$\beta_i = \frac{\mu_{Z_i}}{\sigma_{Z_i}} \quad (i = 1, 2, \ldots, n) \tag{7.28}$$

is called the reliability index.

When random variables A_{Li} are dependent, they can be transferred into a series of independent random variables according to correlation matrix cov(A_{Li}, A_{Lk}) (Ang and Tang 1984). Thus, for large-scale water supply networks, Eq. (7.27) can give the seismic functional reliability of different nodes.

In the above calculations, $g_i(\mu_1, \mu_2, \ldots, \mu_n)$ in Eq. (7.25) can be given by substituting the mean of random variables, $\overline{A_{Lj}}$, in hydraulic equation and solving the deterministic hydraulic equation. Thus, μ_{Z_i} is easy to be obtained. However, as a_j in Eq. (7.24) is difficult to calculate directly, a special strategy policy should be used to get σ_{Z_i}.

Generally, $H_{i\min}$ is a prescribed constant. Therefore the standard deviation of Z_i is equal to that of H_i. That is

$$\sigma_{Z_i} = \sigma_{H_i} \tag{7.29}$$

Thus, the problem becomes how to give the standard deviation of the nodal head H_i.

For water supply network with leakage, the leakage area A_L and nodal head H_i are both random variables. When Eq. (7.17) is expanded at the mean of random variables by Taylor's expansion and only the linear part is preserved, there exists:

$$f(\overline{\mathbf{H}}) + \left.\frac{\partial \mathbf{G}}{\partial \mathbf{H}}\right|_{\mathbf{H}=\overline{\mathbf{H}}} \Delta\mathbf{H} - \left.\frac{\partial \mathbf{Q}_\mathrm{L}}{\partial \mathbf{H}}\right|_{\mathbf{H}=\overline{\mathbf{H}}} \Delta\mathbf{H} - \left.\frac{\partial \mathbf{Q}_\mathrm{L}}{\partial \mathbf{A}_\mathrm{L}}\right|_{\mathbf{A}=\overline{\mathbf{A}}_\mathrm{L}} \Delta\mathbf{A}_\mathrm{L} \approx 0 \tag{7.30}$$

where

$$G = \mathbf{A}\mathbf{R}\left(\mathbf{A}^\mathrm{T}\mathbf{H}\right)^\alpha \tag{7.31}$$

$$\Delta\mathbf{H} = \mathbf{H} - \overline{\mathbf{H}} \tag{7.32}$$

$$\Delta\mathbf{A}_\mathrm{L} = \mathbf{A}_\mathrm{L} - \overline{\mathbf{A}_\mathrm{L}} \tag{7.33}$$

$$\mathbf{A}_\mathrm{L} = (A_{L1}, A_{L2}, \ldots, A_{Lm}) \tag{7.34}$$

Apparently, the hydraulic equation is balanced at the mean of random variables, i.e. $f\left(\overline{\mathbf{H}}\right) = 0$. Let

$$\mathbf{C}_H = \left.\frac{\partial \mathbf{G}}{\partial \mathbf{H}}\right|_{\mathbf{H}=\overline{\mathbf{H}}} - \left.\frac{\partial \mathbf{Q}_\mathrm{L}}{\partial \mathbf{H}}\right|_{\mathbf{H}=\overline{\mathbf{H}}} \tag{7.35}$$

$$\mathbf{C}_\mathrm{L} = \left.\frac{\partial \mathbf{Q}_\mathrm{L}}{\partial \mathbf{A}_\mathrm{L}}\right|_{\mathbf{A}=\overline{\mathbf{A}}_\mathrm{L}} \tag{7.36}$$

Noticing Eqs. (7.32) and (7.33), the Eq. (7.30) will be changed as follows:

$$\mathbf{H} = \overline{\mathbf{H}} + \mathbf{C}_\mathrm{H}^{-1}\mathbf{C}_\mathrm{L}\left(\mathbf{A}_\mathrm{L} - \overline{\mathbf{A}_\mathrm{L}}\right) = \hat{\mathbf{H}} + \mathbf{C}\mathbf{A}_\mathrm{L} \tag{7.37}$$

where

$$\hat{\mathbf{H}} = \overline{\mathbf{H}} + \mathbf{C}_\mathrm{H}^{-1}\mathbf{C}_\mathrm{L}\overline{\mathbf{A}_\mathrm{L}} \tag{7.38}$$

$$\mathbf{C} = \mathbf{C}_\mathrm{H}^{-1}\mathbf{C}_\mathrm{L} \tag{7.39}$$

Equation (7.37) indicates that the relationship between nodal head and leakage area becomes a linear function. Thus, using covariance transfer function formula, there is

$$\sigma_{H_i}^2 = \sum_{j=1}^{m}\sum_{k=1}^{m} c_{ij}c_{jk}\mathrm{cov}\left(A_{Lj}, A_{Lk}\right) \quad (i = 1, 2, \ldots, n) \tag{7.40}$$

where c_{ij} is the element of matrix $\mathbf{C}$ and $\mathrm{cov}\left(A_{Lj}, A_{Lk}\right)$ is the covariance of random variables A_{Lj} and A_{Lk}.

When all leakage areas are independent, there exists:

$$\sigma_{H_i} = \sqrt{\sum_{j=1}^{m} c_j^2\sigma_j^2} \quad (i = 1, 2, \ldots, n) \tag{7.41}$$

where c_j is the diagonal element of $\mathbf{C}$ and σ_j is the standard deviation of leakage area i.

After giving μ_{Z_i} and σ_{Z_i}, the nodal reliability index can be given.

The advantage of the above method is that the nodal reliability index can be given by performing hydraulic analysis only once. Using the method, the nodal reliability index of the water distribution networks in Fig. 7.2 is shown in Fig. 7.3.

As a practical example, Fig. 7.4 shows an actual water supply network with 13 water supply factories, 463 nodes and 977 pipes. Using the above analysis method, the water supply network is analyzed considering the seismic intensity of VIII. The results of partial node reliability index are shown in Table 7.2.

Table 7.3 gives a comparison of the connectivity reliability and the functional reliability of the network under the same earthquake inputs. Obviously, the connectivity reliability and the functional reliability of the network are totally different and the later is much lower than the former.

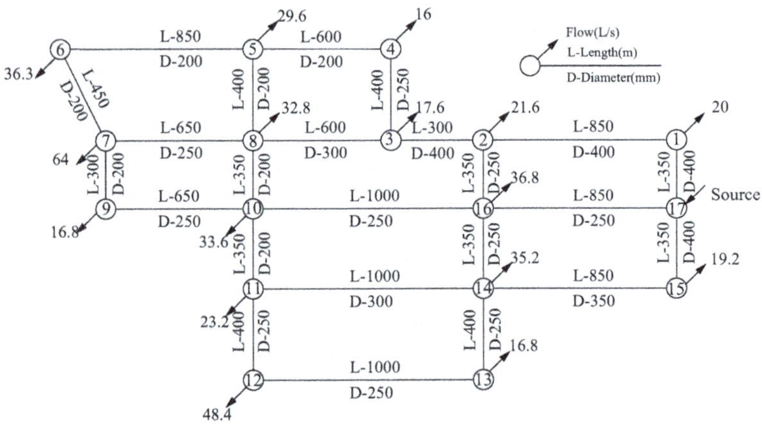

Fig. 7.2 A water supply network

Fig. 7.3 Reliability index of different nodes

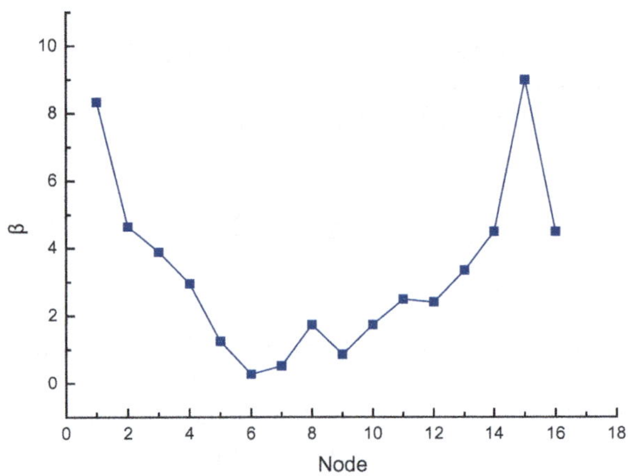

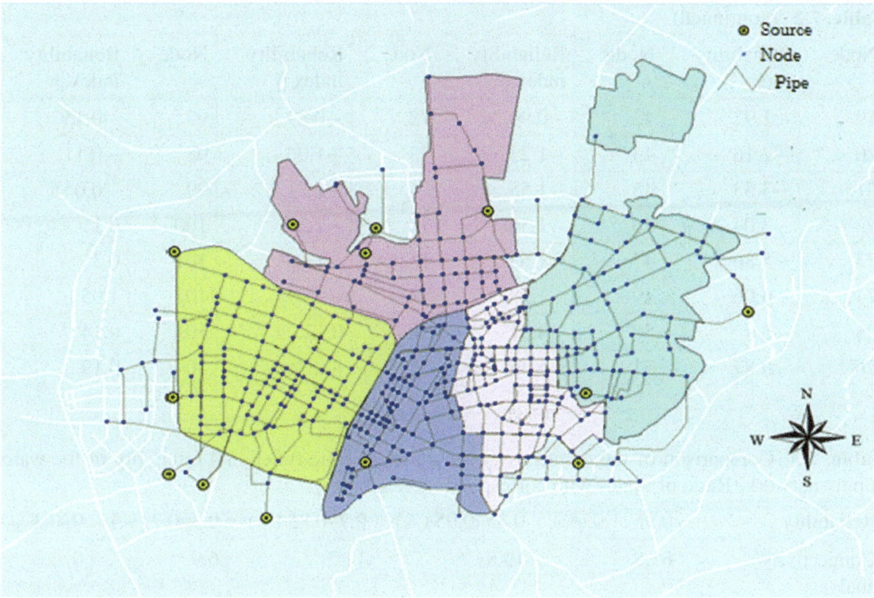

Fig. 7.4 A real water supply network

Table. 7.2 Node reliability index of a water supply network

Node	Reliability index β	Node	Reliability index β	Node	Reliability index β	Node	Reliability index β
2	1.65	26	−0.87	52	−0.51	80	−1.34
3	−0.89	27	−0.89	53	−1.25	81	−0.48
4	−0.77	28	−0.52	54	−1.33	82	−0.11
5	−0.64	29	−2.11	55	−2.14	83	−10.0
6	−0.6	30	−1.03	57	−4.12	84	−10.0
7	−0.82	31	−28.72	58	−5.06	85	−10.0
8	−0.98	32	−0.99	59	3.82	86	1.6
9	−2.03	33	−0.97	61	−0.73	88	1.48
10	−0.03	34	−1.35	62	−0.91	89	−10.0
11	−1.34	35	−0.77	63	−1.25	90	−0.17
13	−0.96	36	−2.43	64	−1.45	91	1.95
14	−1.37	37	−2.54	65	−1.5	92	1.95
15	0.53	38	−3.18	66	0.39	93	0.67
16	−0.87	39	1.77	67	−2.03	94	0.06
17	−1.64	40	2.94	68	−4.31	95	−0.28
18	−2.38	41	−1.04	71	−0.1	96	−0.17

(continued)

Table. 7.2 (continued)

Node	Reliability index β	Node	Reliability index β	Node	Reliability index β	Node	Reliability index β
19	−1.92	42	−0.96	72	−0.89	97	−0.15
20	−2.16	43	−1.23	73	−1.06	98	−0.11
21	−3.33	45	−1.55	74	−1.21	99	−0.05
22	−3.04	46	−1.56	75	−1.29	100	0.1
23	−5.86	47	−1.63	76	−1.06	101	0.3
24	−6.06	49	−1.58	77	−1.45	102	1.05
25	−6.7	50	−4.61	78	0.13	103	0.14
26	−0.87	51	−5.35	79	−0.71	104	0.19

Table. 7.3 Comparison of the connectivity reliability and the functional reliability of the water supply network (Ratio of nodes with corresponding reliability)

Reliability	0.95–1.0 (%)	0.85–0.95 (%)	0.7–0.85 (%)	0.3–0.7 (%)	0–0.3
Connectivity analysis	61.8	19.8	11.5	6.9	0
Functional analysis	17.56	2.67	2.89	5.32	71.56%

References

Ang, A. H. S., & Tang, W. (1984). *Probability concepts in engineering planning and design*. Wiley.

Chen, L. L., & Li, J. (2003). Leakage analysis of water supplying networks. *Earthquake Engineering and Engineering Vibration, 23*(1), 115–121 (in Chinese).

Yan, X. S., & Zhao, H. B. (1986). *Theory and computation of water supply network*. China Architecture and Building Press (in Chinese).

Chapter 8
Aseismic Optimal Design of Lifeline Networks

8.1 Introduction

The goal of seismic reliability analysis of lifeline networks is not just to evaluate the possibility of failure of the lifeline systems subjected to earthquakes. The more important target is to provide a suitable tool for the aseismic design of lifeline networks. In the design process of a lifeline network, the basic consideration is the system component reliability (or structural reliability) and the topological structure of the network. Obviously, the design of network topology is a more important consideration in the system level. For a lifeline network, modifying the topology of network by adding some elements or removing some elements is a feasible way to improve system seismic reliability. Therefore, combined with a suitable system component design, the topological optimization is important in the design of lifeline networks.

As shown in Chaps. 6 and 7, there are two kinds of reliability analysis methods. Correspondingly, there are two kinds of optimization design methods: optimization based on connectivity reliability and optimization based on functional reliability. The basic optimization technology is based on the modern combinatiorial optimization theory, such as the genetic algorithm (Holland 1975), the simulated annealing algorithm (Kirkpatrick et al. 1983; Cerny 1985), and the particle swarm algorithm (Kennedy and Eberhart 1995), etc. In principle, all these algorithms can be employed to solve the above two kinds of optimization problems. For simplicity, only the genetic algorithm is introduced combined with different targets in this chapter.

J. Li and W. Liu, *Lifeline Engineering Systems*,
https://doi.org/10.1007/978-981-15-9101-3_8

8.2 Network Topology Optimization Based on Connectivity Reliability

8.2.1 Topology Optimization Model

To carry out an optimization design of a network is actually to find an optimal network topology that satisfies nodal seismic reliability constraints. The optimization model based on the connectivity reliability can be mathematically formulated in the following form:

$$\begin{aligned} &\text{minimize} \quad C\left(G^{*}\right)=\sum \gamma_{ij} c_{ij} \\ &\text{subject to} \quad P_k \geq P_0 \quad k=1,2,\ldots,n \end{aligned} \tag{8.1}$$

where G^* is a possible solution of G; G denotes a graph of a network; γ_{ij} is a connectivity factor which takes a value of 0 if no edge exists between node i and node j and otherwise 1; P_k is the seismic connectivity reliability between terminal k and the source, which can be given by using the RDA technology in Chap. 7; P_0 is the reliability threshold of the system; n is the number of the sinks; c_{ij} is the cost of the pipe between nodes i and j which can be evaluated by the following empirical formula:

$$c_{ij}=(ad_{ij}-b)l_{ij} \tag{8.2}$$

where a and b are empirical coefficients, d_{ij} is the diameter and l_{ij} is the length of pipe between node i and j.

Several combinational optimization methods have been proposed for solving the above network optimization problem, such as the genetic algorithm (Li et al. 2008), the particle swarm algorithm (Liu et al. 2012), and others (Liu et al. 2014). Here, only the genetic algorithm is described in detail as follow.

8.2.2 Genetic Algorithm

Genetic algorithm (GA) was proposed by Holland (1975) in the field of machine learning. In the network topology optimization, a generation including many chromosomes is created initially where each chromosome represents a network. Then by using the genetic operators including selection, crossover and mutation operators, a new generation is evolved. After a number of iterations or when some specified criteria are met, a feasible optimal solution can be reached. In the following subsections, the method is introduced in cooperation with topology optimization based on connectivity reliability of networks.

Fig. 8.1 A bridge network

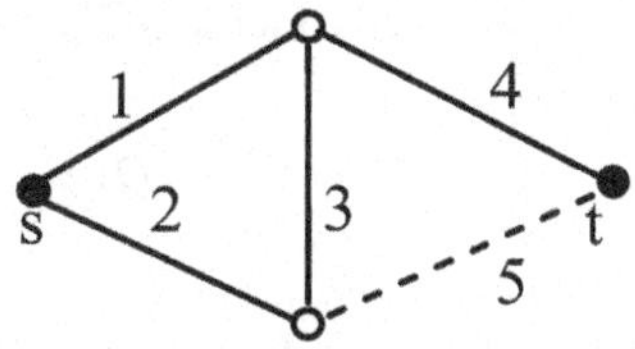

(1) **Representation of a network**

In order to operate a GA for network optimization, it is required that the network is represented properly. Define a network $G = (g_1, g_2, \cdots, g_n)$ as a chromosome, where g_n is a gene. When $g_i = 1$, it means that the gene possesses a corresponding edge of *G*, while '0′ means not. For example, a bridge network shown in Fig. 8.1 can be represented as 11110. A certain number of chromosomes consist of a population of GA.

(2) **Generation of initial populations**

It is reasonable to hope the first-generation population contains as many diverse genes as possible. Therefore, generating a series of random network graphs is an important job in this process. To generate a random graph, all the genes in a chromosome is initialized with '0′s. Then each gene in the chromosome is changed to '1′ randomly with a prescribed probability. The probability could be determined by a specified seismic reliability threshold P_0.

It should be noted that not all initial graphs generated by the above method are feasible. Before taking a generated graph as a chromosome, it is necessary to judge the connectivity of each graph in the initial generation and modify the disconnected graphs by adding necessary edges.

(3) **Evaluation of individuals**

A chromosome is also called an individual. For each generation, all individuals need to be evaluated to determine their fitness. The individual with larger fitness has more chance to be transferred to the next generation.

The fitness function can be defined as:

$$F(X) = M - C(X) \tag{8.3}$$

where *M* is a large value and *C*(*X*) is the network cost of corresponding individual *X*.

Obviously, not all individuals are feasible solution because some of them may not satisfy the constraint condition in Eq. (8.1). Therefore, a penalty factor should be introduced to the unfeasible solutions. That is

$$F(X) = \begin{cases} M - C(X) & P_{\min} \geq P_0 \\ M - C(X) - Q(X) & P_{\min} < P_0 \end{cases} \tag{8.4}$$

The penalty factor $Q(X)$ can be defined as follows:

$$Q(X) = (C_{\max} - C_{\min})\frac{P_{\max} - P(X)}{P_{\max} - P_{\min}} \tag{8.5}$$

where $P_{\max}$ and $P_{\min}$ denote the maximum and minimum connectivity reliabilities in one generation; $P(X)$ denotes connectivity reliability of the individual; $C_{\max}$ and $C_{\min}$ denote the maximum and minimum individual cost in one generation, respectively.

(4) **Selection operator**

The selection operator is executed to select individuals with higher fitness into the next generation (inheritance). Although many selection operators are available, the elitist selection operators and the roulette wheel selection are adopted here. The operating process of selection operator is.

① Select the individual with the highest fitness into next generation and remove it from current generation;
② Determine the selection probability of individual i

$$\mathrm{P}_i = F(i) / \sum_{j=1}^{\mathrm{n}} F(j) \tag{8.6}$$

where $F(i)$ is the fitness of individual i and n is the number of remaining genes in the current generation;
③ Let $P_0 = 0$ and determine the location of individual i in $\sum_{j=0}^{\mathrm{i}-1} P_j, \sum_{j=1}^{\mathrm{i}} P_j)$;
④ Produce a random number between 0 and 1 and judge the numerical range it belongs to. Then the individual corresponding to the numerical range is transferred to the next generation and removed it from current generation;
⑤ Repeat ②–④ until enough individuals in the current generation are transferred to the next generation.

In the above process, step 1 corresponds to the elitist selection operator and the other four steps correspond to the roulette wheel selection operator.

(5) **Crossover operator**

Hybridization is an important link in the process of biological evolution. To simulate this link, in the genetic algorithm, the crossover operator is introduced to guarantee the global searching capability of the algorithm. This operator takes two randomly selected individuals and randomly exchange some parts of genes. That is:

$$C(X_i, X_j) = C\left\{\begin{array}{l} X_i = g_1^i, g_1^i, \ldots, g_{l-1}^i, \underline{g_l^i, \ldots, g_m^i}, g_{m+1}^i, \ldots, g_n^i \\ X_j = g_1^j, g_1^j, \ldots, g_{l-1}^j, \underline{\underline{g_l^j, \ldots, g_m^j}}, g_{m+1}^j, \ldots, g_n^j \end{array}\right\}$$

$$= \left\{ \begin{array}{l} X_i^{'} = g_1^i, g_1^i, \ldots, g_{l-1}^i, \underline{g_l^j, \ldots, g_m^j}, g_{m+1}^i, \ldots, g_n^i \\ X_j^{'} = g_1^j, g_1^j, \ldots, g_{l-1}^j, \underline{\underline{g_l^i, \ldots, g_m^i}}, g_{m+1}^j, \ldots, g_n^j \end{array} \right\} \tag{8.7}$$

If the new individual is a disconnected graph, the individual should be modified by adding necessary edges.

In the crossover operation, the selected number m of individuals is

$$m = P_c n \tag{8.8}$$

where n is the number of individuals in the current generation; P_c is the crossover probability. Generally, $P_c = 0.6 \sim 0.8$.

(6) **Mutation operator**

Variation is another part of biological evolution. To simulate this part, in the genetic algorithm, the mutation operator is used to guarantee the local searching capability of the algorithm. The process of this operator is to randomly reverse the gene in individuals at a prescribed probability. That is

$$\mathrm{M}\left(X_i = g_1^i, g_2^i, \ldots, g_l^i, \ldots, g_n^i,\right) = \left(X_i^{'} = g_1^i, g_2^i, \ldots, \bar{g}_l^i, \ldots, g_n^i,\right) \tag{8.9}$$

where $\bar{g}$ is a reverse event of g.

The mutation probability can be chosen as 0.01–0.1.

(7) **Stopping criteria**

After executing the selection, crossover and mutation operators, a new generation is produced. For the networks in the new generation, connectivity reliability analysis is performed for each individual to give the fitness, penalty function, etc. Then a new round of selection, crossover and mutation operators are applied to this generation and yield the next generation. The process repeats and the stop criteria for the process are:

$$GR_{\mathrm{N}} > GR_{\max} \tag{8.10}$$

or

$$\delta = \frac{\sigma(f_j, j = 1, 2, \ldots, n_N)}{\mu(f_j, j = 1, 2, \ldots, n_N)} \le \varepsilon \tag{8.11}$$

where GR_{N} is the number of the generations, $GR_{\max}$ is the prescribed number of the generations, σ and μ are the standard deviation and the mean of the individual finesses in a generation.

8.2.3 Examples

【Example 1】 Figure 8.3 is a simple gas network with 10 nodes. The network consists of 14 pipes and the seismic reliability of each pipe is 0.9. The initial cost of the network is \$7.69 million.

By using the GA method, the optimal solutions which, respectively, satisfy the reliability constraints of 0.7 and 0.9 are shown in Figs. 8.4 and 8.5, respectively. The total cost with the lowest nodal reliability is shown in Table 8.1. Obviously, the optimization of the network is effective (Fig. 8.2).

Table. 8.1 Optimization results for the network in Fig. 8.1

Reliability constraint P_0	The minimal nodal reliability	Cost
0.7	0.747	\$4.46 million
0.9	0.920	\$5.26 million

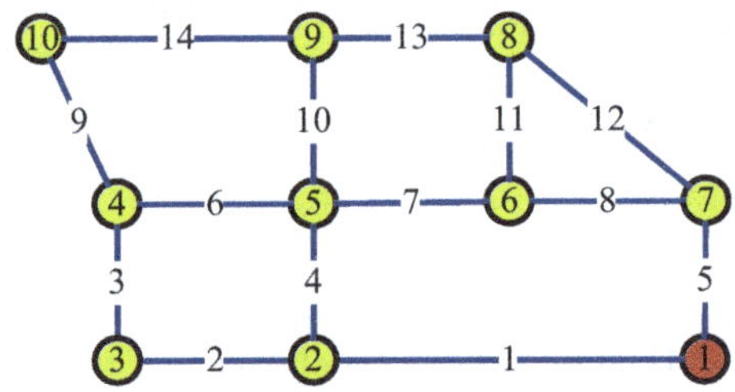

Fig. 8.2 A simple network

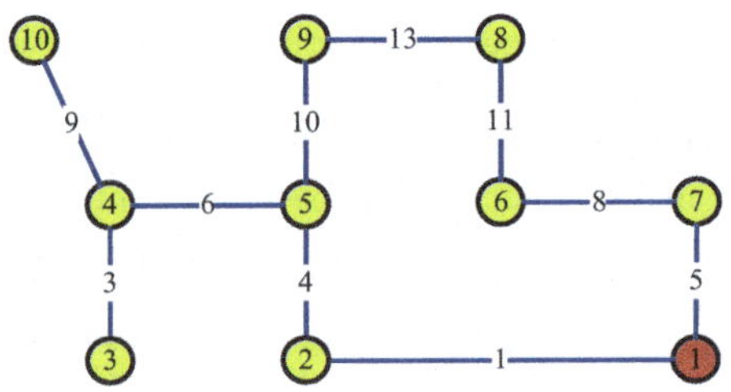

Fig. 8.3 The optimal network when P_0 is 0.7

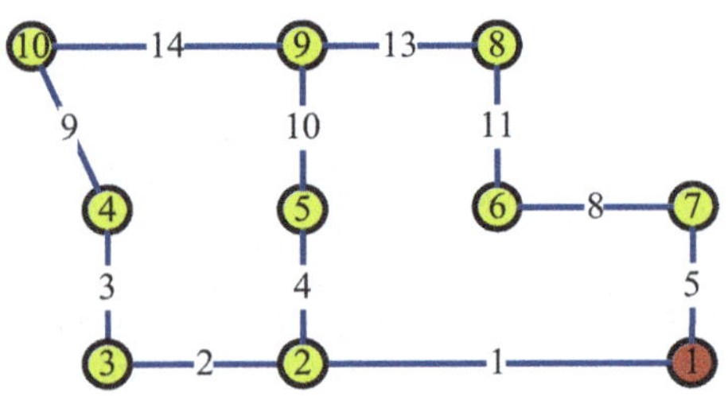

Fig. 8.4 The optimal network when P_0 is 0.9

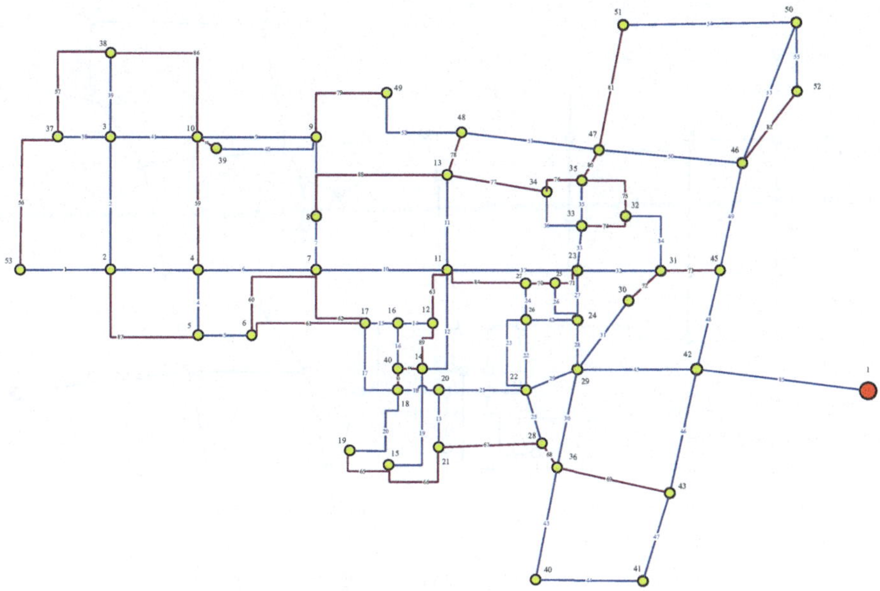

Fig. 8.5 Initial design scheme of the gas supply network in Puyang City

【Example 2】 An actual gas supply network in Puyang City, located in China, is investigated to show the validity of the above network topology optimization algorithm.

Based on the connectivity reliability analysis, the seismic reliability of gas supply network in Puyang City is very low and the lowest connectivity reliability is only 0.389 for the seismic intensity of VIII. Therefore, an initial retrofitting design scheme is given Fig. 8.5, which consists of 52 nodes and 89 pipelines.

By adopting the above network topology optimization algorithm, the optimal networks with different minimal reliability, 0.7, 0.8 and 0.9, respectively, are derived. In the computation, the number of chromosomes in each generation is 90 and the number of generations is 50. An optimal network is shown in Fig. 8.6, in which reliability constrains is 0.8 and the seismic intensity is VIII. Table 8.2 shows the cost of optimal network with different reliability constraints.

8.3 Topology Optimization of Water Supply Network

8.3.1 Optimization Model

For water supply systems, the optimization based on seismic functional reliability should be adopted. In this kind of optimization, not only the reliability constraints

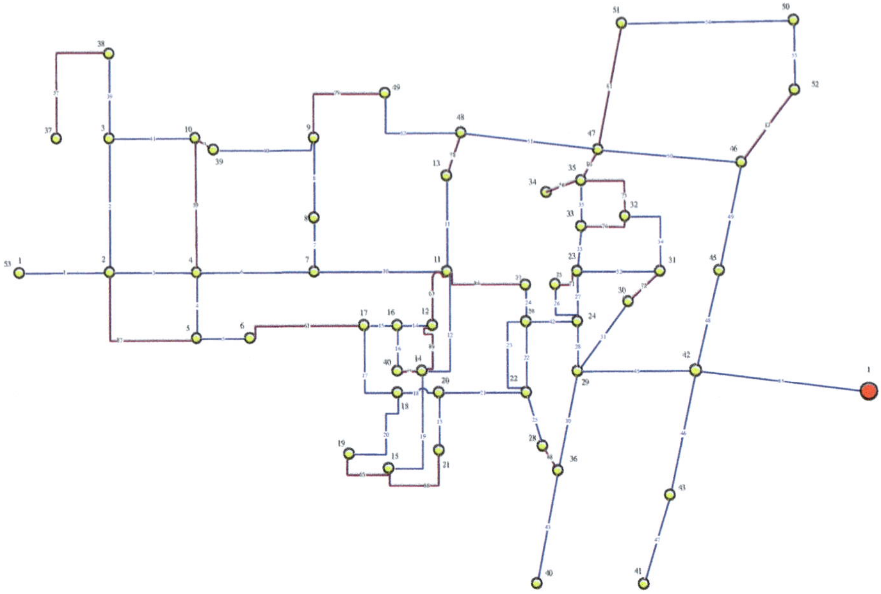

Fig. 8.6 The optimal design scheme of the gas supply network in Puyang City

Table. 8.2 Cost of optimal network

Reliability Constraints	0.7	0.8	0.9
Cost (¥ Million Yuan)	27.33	28.13	31.93

should be considered but also the operation function constraints should be considered as well. Generally, the later can be considered by the constraints of the economic water velocity and pipe diameter constraints. Therefore, the topology optimization model for water supply networks can be written as follows:

$$\min \quad \mathrm{C} = \sum \gamma_{ij} c_{ij}$$

$$\text{s.t.} \quad \begin{cases} \beta_j \geq \beta_0 & j = 1, 2, \ldots, n \\ V_{\min} \leq V_i \leq V_{\max} & i = 1, 2, \ldots, n \\ d_{\min} \leq d_i \leq d_{\max} & i = 1, 2, \ldots, n \end{cases} \tag{8.12}$$

where β_j is the seismic functional reliability index of node j which can be calculated by using the method in Chap. 7; n is the number of nodes in the network; β_0 is the threshold of the functional reliability index; $V_i$is the velocity of water flow in *ith* pipe of the network, $V_{\max}$ and $V_{\min}$ are the maximum and minimum value of economic velocity, respectively; d_i is pipe diameter of *ith* pipe in the network, $d_{\max}$ and $d_{\min}$ are the specified maximum and minimum value of pipe diameter, respectively.

8.3.2 Algorithms for Seismic Topology Optimization

Obviously, the genetic algorithm described in the above section can also be adopted to solve the above optimization problem. However, considering different background, there are some different treatment in the process of optimization which can be described as follows.

(1) **Coding**

An integer coding should be adopted in the topology optimization of water supply networks. That is, each pipe in the network is represented by an integer ranging from 0 to n where 0 indicates that the pipe is not placed and other integers indicate that the pipe with a specific diameter is laid. For example, if the possible diameters of a pipe are 100, 150 and 200 mm, respectively, the pipe can be represented by an integer ranging from 0 to 3, where 0 indicates that the pipe is not placed and 1, 2 and 3 indicate that the pipe with diameter of 100, 150, and 200 mm is laid, respectively.

(2) **Generating initial population**

To generate a random network, all the genes in the graph is initialized with '0′s'. Then each edge in the graph are chosen in turn and changed to an integer ranging from 1 to n according to a prescribed probability. As a result, a chromosome is generated. For the disconnected graph, the necessary edges are added. Generally, the first generation should include enough individuals (chromosomes). Shown in Fig. 8.7 are two examples of initial individuals.

(3) **Evaluation of individuals**

For the networks in a generation, their fitness should be evaluated. The fitness function can be expressed as follows:

$$F(s) = M - C(X) - P(X) \tag{8.13}$$

where M is a large number, $C(X)$ denotes the cost of Xth network, $P(X)$ denotes the penalty factor which can be written as follows:

$$P(X) = \begin{cases} (C_{\max} - C_{\min}) \frac{\Delta\beta(X) - \Delta\beta_{\min}}{\Delta\beta_{\max} - \Delta\beta_{\min}} & \beta(X) < \beta_0 \\ 0 & \beta(X) \geq \beta_0 \end{cases} \tag{8.14}$$

where $C_{\max}$ and $C_{\min}$ denote the maximum and minimum cost in one generation, respectively. $\Delta\beta_{\max}$ and $\Delta\beta_{\min}$ are the maximum and minimum of all $\Delta\beta(X)$ in one generation, respectively.

$$\Delta\beta(X) = \sum_{i=1}^{n} (\beta_0 - \beta_i) \tag{8.15}$$

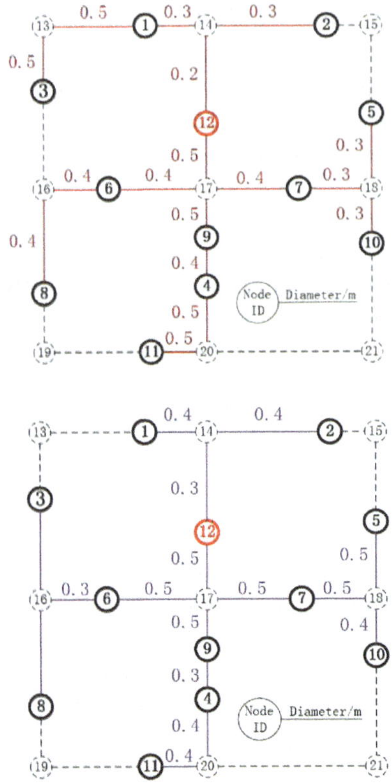

$$Y_1 = \begin{pmatrix} 0 & 3 & 5 & 0 & 0 & 0 & 0 & 0 & 0 & 0 & 0 & 2 \\ 0 & 0 & 0 & 0 & 0 & 0 & 0 & 0 & 0 & 0 & 0 & 0 \\ 0 & 0 & 0 & 0 & 0 & 0 & 0 & 0 & 0 & 0 & 0 & 0 \\ 0 & 0 & 0 & 0 & 0 & 0 & 0 & 0 & 4 & 0 & 5 & 0 \\ 0 & 0 & 0 & 0 & 0 & 0 & 3 & 0 & 0 & 3 & 0 & 0 \\ 0 & 0 & 0 & 0 & 0 & 0 & 4 & 4 & 0 & 0 & 0 & 0 \\ 0 & 0 & 0 & 0 & 0 & 0 & 0 & 0 & 0 & 0 & 0 & 0 \\ 0 & 0 & 0 & 0 & 0 & 0 & 0 & 0 & 0 & 0 & 0 & 0 \\ 0 & 0 & 0 & 0 & 0 & 0 & 0 & 0 & 0 & 0 & 0 & 5 \\ 0 & 0 & 0 & 0 & 0 & 0 & 0 & 0 & 0 & 0 & 0 & 0 \\ 0 & 0 & 0 & 0 & 0 & 0 & 0 & 0 & 0 & 0 & 0 & 0 \\ 0 & 0 & 0 & 0 & 0 & 0 & 0 & 0 & 0 & 0 & 0 & 0 \end{pmatrix}$$

$$Y_2 = \begin{pmatrix} 0 & 4 & 0 & 0 & 0 & 0 & 0 & 0 & 0 & 0 & 0 & 0 \\ 0 & 0 & 0 & 0 & 0 & 0 & 0 & 0 & 0 & 0 & 0 & 3 \\ 0 & 0 & 0 & 0 & 0 & 5 & 0 & 0 & 0 & 0 & 0 & 0 \\ 0 & 0 & 0 & 0 & 0 & 0 & 0 & 0 & 3 & 0 & 4 & 0 \\ 0 & 0 & 0 & 0 & 0 & 0 & 5 & 0 & 0 & 0 & 0 & 0 \\ 0 & 0 & 0 & 0 & 0 & 0 & 0 & 3 & 0 & 0 & 0 & 5 \\ 0 & 0 & 0 & 0 & 0 & 0 & 0 & 0 & 5 & 4 & 0 & 0 \\ 0 & 0 & 0 & 0 & 0 & 0 & 0 & 0 & 0 & 0 & 0 & 0 \\ 0 & 0 & 0 & 0 & 0 & 0 & 0 & 0 & 0 & 0 & 0 & 5 \\ 0 & 0 & 0 & 0 & 0 & 0 & 0 & 0 & 0 & 0 & 0 & 0 \\ 0 & 0 & 0 & 0 & 0 & 0 & 0 & 0 & 0 & 0 & 0 & 0 \\ 0 & 0 & 0 & 0 & 0 & 0 & 0 & 0 & 0 & 0 & 0 & 0 \end{pmatrix}$$

Fig. 8.7 Two initial individuals

where n is the node number in the network, and β_0 is a specified system functional reliability index.

Then the following three operators, i.e. the selection, crossover and mutation operators, are executed to generate the next generation which is similar to those in the GA for topological optimization based on connectivity reliability. Finally, the stopping criteria of the algorithm can be taken as Eq. (8.10) or (8.11).

8.3.3 Examples

【Example 1】 A water supply network in Fig. 8.8 is adopted as an example to confirm the applicability of the above optimization algorithms. In the network, there are five alternative diameters, 200, 250, 300, 350 and 400 mm. The seismic intensity is VIII and the reliability index constraints is $\beta_0 = 0.53$, i.e. the corresponding minimum nodal reliabilities are 0.7. By using the above optimization algorithm, the best results are shown in Fig. 8.9.

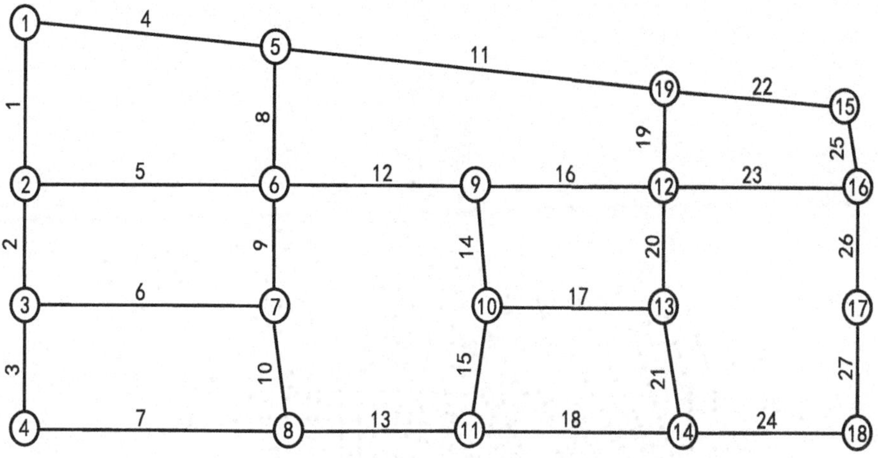

Fig. 8.8 A water supply network

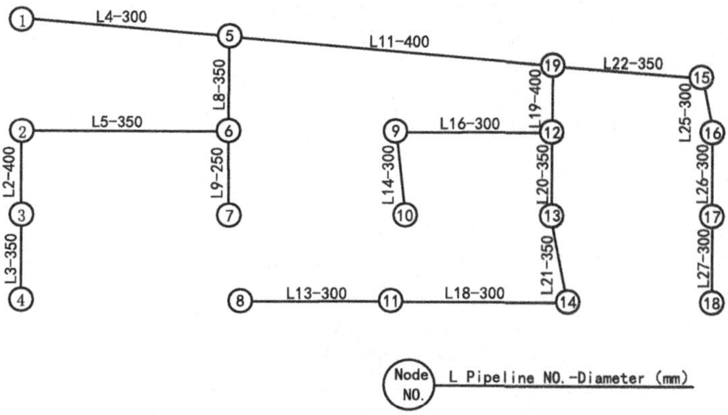

Fig. 8.9 Best network when $\beta_0 = 0.53$

【**Example 2**】 Another optimal design example for a practical engineering network is the aseismic design for the water supply network of Shenyang city, which is located in the northeast part of China. Figure 8.10 shows the location of the water supply points in the city. Taking the seismic reliability index as $\beta_0 = 1.58$, which means the lowest functional reliability of the network is 0.94, the optimal network is derived as shown in Fig. 8.11. The optimal result consists of 726 pipelines and part of them are shown in Table 8.3. During the calculation, the number of each generation is 50, and the crossover probability is 0.7.

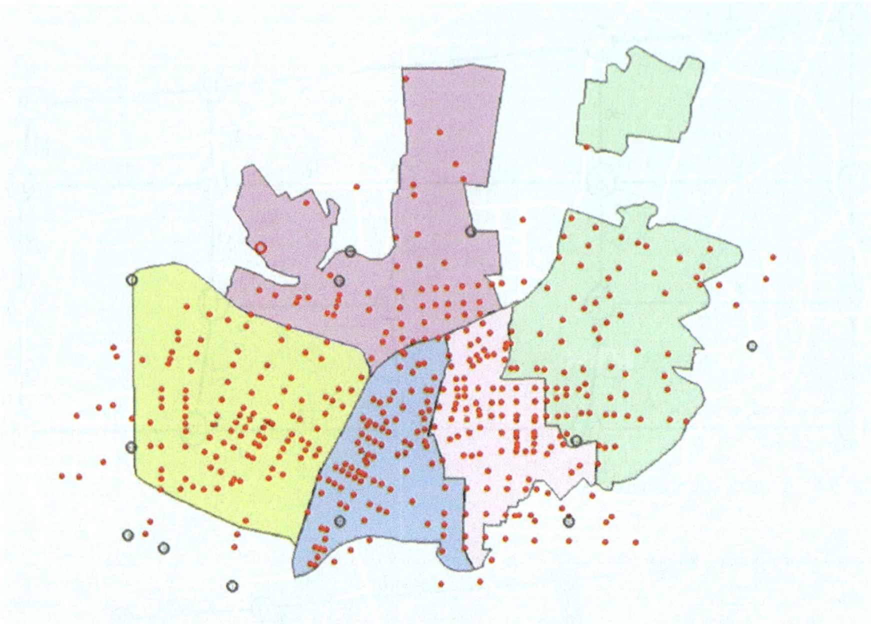

Fig. 8.10 Location of the water demand points in Shenyang City

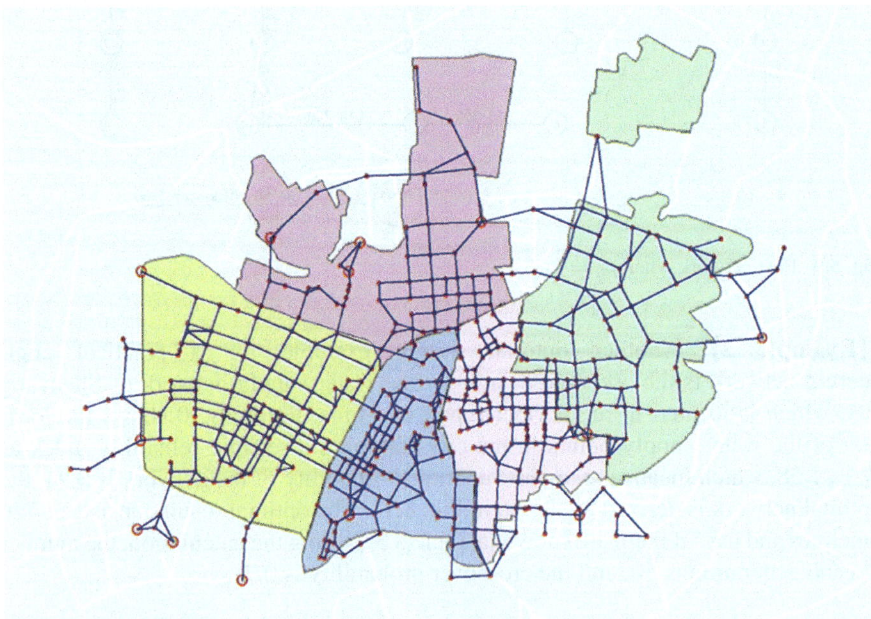

Fig. 8.11 The optimal water distribution network of Shenyang City

Table. 8.3 Parts of design results

No	Length (m)	Diameter (mm)	No	Length (m)	Diameter (mm)
1	689.93	900	46	350.21	400
2	773.5	1100	47	142.21	400
3	541.18	400	48	753.03	400
16	690.79	750	49	736.66	400
17	212.39	600	50	277.05	550
20	407.43	600	52	590.44	600
25	389.33	450	647	1082.06	500

References

Černý, V. (1985). Thermodynamical approach to the traveling salesman problem: An efficient simulation algorithm. *Journal of Optimization Theory and Applications, 45*(1), 41–51.

Holland, J. H. (1975). Adaptation in neural and artificial systems. University of Michigan Press.

Kennedy, J. Eberhart, R. (1995). Particle swarm optimization. In *IEEE international conference on neural networks*. Perth, Western Australia: Institute of Electrical Electronics Enginee, pp. 1492–1948.

Kirkpatrick, S., Gelatt, C. D., & Vecchi, M. P. (1983). Optimization by simulated annealing. *Science, 220,* 671–680.

Li, J., Liu, W., & Bao, Y. F. (2008). Genetic algorithm for seismic topology optimization of lifeline network systems. *Earthquake Engineering and Structural Dynamics, 37*(11), 1295–1312.

Liu, W., Xu, L., & Li, J. (2012). Algorithms for seismic topology optimization of water distribution network. *Science China: Technological Sciences, 55*(11), 3047–3056.

Chapter 9
Simulation and Control of Composite Lifeline System

9.1 Introduction

Since the San Fernando earthquake in 1971, the investigations on almost all catastrophic earthquakes which occurred near cities indicates that different lifeline engineering systems are interacted with each other during disasters. Table 9.1 listed the damage of lifeline systems and their interactions during several typical catastrophic earthquakes.

These facts mean that, for modern society, the function of lifeline engineering systems is interrelated. Several interactional lifeline engineering systems can be called composite lifeline system. The theory for seismic response simulation and control of composite lifeline system, which studies the performance evolution of lifeline engineering systems during disaster and post-disaster recovery process, is the theoretical basis for urban disaster prevention planning, urban lifeline engineering retrofit planning and resilience improvement. Actually, this kind of research is a new milestone in lifeline earthquake engineering (Li, 1992).

The branch system of a composite lifeline system may be different according to different research targets. For example, when considering the interactions of lifeline engineering systems, two or more lifeline engineering systems can be chosen to construct a composite lifeline system, such as power electricity systems and water supply networks; power and gas supply systems; firefighting, transportation and communication systems, etc. When considering the economic loss of composite lifeline system, the systems with obvious supply and demand relationship should be selected, such as power electricity systems, water supply networks and important industry systems, etc. Furthermore, when considering the comprehensive disaster control of a modern city, for example, the secondary fire control, then the firefighting systems, transportation systems, water supply networks and communication systems should be included. By considering the interactions between different lifeline systems, the total disaster resilience level of a city can be identified. In other words, the evaluation of disaster control capability reflects the total disaster prevention capability of a city.

J. Li and W. Liu, *Lifeline Engineering Systems*,
https://doi.org/10.1007/978-981-15-9101-3_9

Table 9.1 Lifeline system damages and their interactions

Earthquake	Water supply system	Gas supply system	Power electricity system
1995 Kobe Earthquake ($M_W = 6.9$)	1.1 million customers encountered water stoppage due to 1610 breaks in water distribution network. A week later, only one-third breaks were repaired. Lack of water brought a serious hamper for firefighting	5190 breaks appeared. Among these breaks, 109 breaks appeared on medium-pressure pipelines. After the earthquake, 857 thousand customers encountered no gas supply and repair lasted 3 months	275KV and 77KV substations suffer serious damages. One million customers encountered electricity cutoff and repair lasted 6 days. Moreover, the recovery of power aggravated the fire disaster
1994 Northridge Earthquake ($M_W = 6.7$)	1400 breaks appeared in water distribution network of Los Angeles, including more than 100 breaks in water mains	Over 150,000 leakages appeared in gas supply network of Los Angeles, resulting in many fire accidents	Many 230KV and 500KV high-voltage substations suffered serious damages. Over 1.1 million customers encountered electricity cutoff
1989 Loma, Prieta Earthquake ($M_W = 6.9$)	Over 350 breaks appeared in water distribution network. Meanwhile, electricity cutoff resulted in serious difficulty in water supply and lack of water a serious hamper for firefighting	More than 1000 leakages appeared in gas supply network and many oil tanks on soft soil ground were destroyed	The serious damages of 230KV and 500KV high-voltage substations resulted in that over 1.4 million customers encountered electricity cutoff

Based on the disaster response simulation of a composite lifeline system, the system control provides a basis for improving the disaster resilience of a city. Actually, for a modern city, not only the disaster prevention capability should be evaluated, more important is to improve the disaster resilience by designing and retrofitting of lifeline systems. The previous content of this book provides the basic technology for the seismic analysis, design and retrofit of single lifeline engineering system. In this chapter, the disaster response analysis of composite lifeline system and urban disaster control will be introduced.

In history, the first author of the book presented the concept of urban seismic disaster field and proposed the hierarchically control method of large system theory to control the disaster response of lifeline systems firstly (Li 1991). In 1993, Li systematically discussed the control of seismic disaster field and presented simulation control models based on the seismic response simulation technology of lifeline systems (Li 1993). In 1996, considering the interactions between different lifeline engineering systems, a basic model for urban seismic disaster field simulation and control was established (Li and Song 1996a, b). In 2000, combined Petri net analysis method with the network connectivity reliability analysis, Han et.al developed a Multi-Resource-Multi-Entity model and used it in the disaster field simulation of Shanghai (Han et al., 2000).

From the viewpoint of interactions between different lifeline systems, Scawthorn (1992) adopted an interaction matrix to identify the influence of water distribution networks on the earthquake secondary fire disaster. In 1994, Kameda gave the interaction matrix of urban main lifeline engineering systems (Kamada, 1994). From 1995 to 1996, taking Memphis city as an example, Shinozuka analyzed the relationships between water distribution networks and power electricity systems using the Monte Carlo simulation method and gave the nodal flow and node head after earthquakes considering the influence of power electricity systems (Shinozuka, 1996). After 2010, more and more researches started to focus the earthquake resilience of modern society and most of such kind of works actually belong to simulation and control of composite lifeline system (Ouyang and Wang, 2015; Nozhati et al., 2019).

9.2 Disaster Response Simulation of Composite Lifeline System

9.2.1 Fundamentals of Discrete Event Dynamic Simulation

Discrete event dynamic system is a system whose state only changes at a series of discrete time points and will keep intact within the time interval. A simple discrete event dynamic system can be illustrated by a single-server queue system in Fig. 9.1. In the figure, the arrival and departure of customers are discrete events. These discrete events change the state of the system.

Generally, the basic characteristics of the discrete event dynamic system can be stated as follows:

(1) The relationship between the system variables cannot be expressed by continuous-time state equations;
(2) The main variables of the system are random variables.

Because of these characteristics, the simulation of the discrete event dynamic system is generally carried out by a system dynamic model and random sampling technology. By the simulation of the system model and statistics of results, the mean tendency of the system response can be investigated.

Classical time domain simulation model can also be illustrated in Fig. 9.1. In this single-server queue system, the basic entities can be divided into flow entity and resource. The arrival time of flow entity (customer) can be realized/obtained by a probabilistic model. The flow entity flows into the system according to prescribed rule and accepts the system service. Meanwhile, the system resource (server) deals with flow entity according to the prescribed rule. The system's state changes and the change are recorded by simulation clock. Generally, the advance of a simulation clock

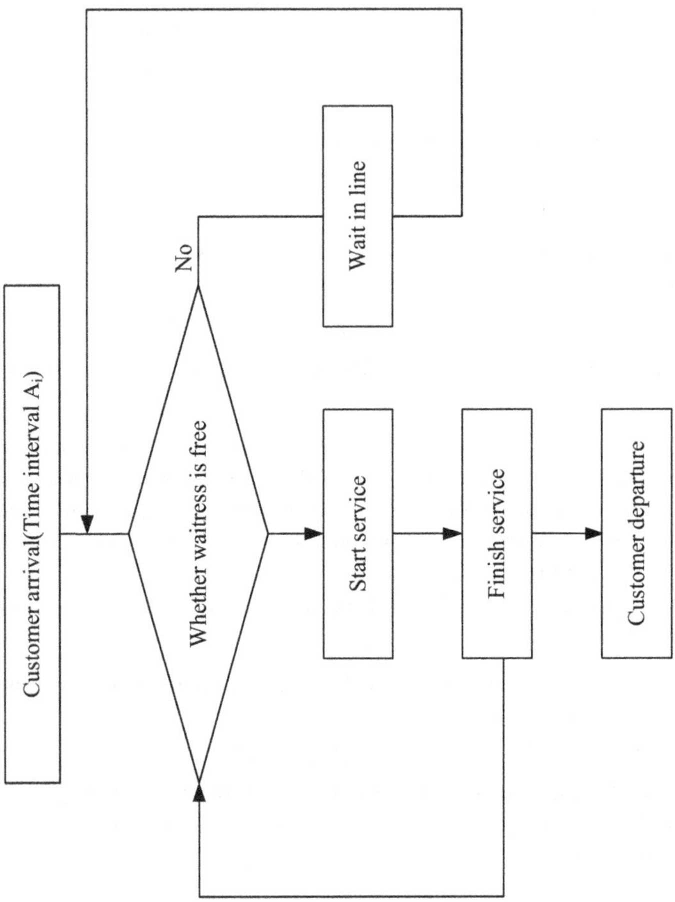

Fig. 9.1 Single-server queue system simulation model

can adopt event scheduling method, i.e., the next earliest event which changes state of the system.

The simulation model is executed in a certain time interval to simulate the occurrence of events and the process of system state change. Then the results, including the mean queue length, server utilization rate and mean stay time of customer, can be analyzed to evaluate the i.e., the system and to provide the technical support for system design and improvement.

From the process mentioned above, the simulation of the discrete event dynamic system consists of three parts, system model establishment, simulation execution and simulation result analysis.

(1) **System modeling**

Based on the occurrence and development of the major events in the system, a flowchart such as Fig. 9.1 for system movement could be established. This logical flowchart illustrates the system simulation model and formulates the foundation of system simulation programs. In order to consider the randomness in the process of event occurrence and development, the probabilistic models for system major parameters should be identified, which could be derived by statistical modeling.

(2) **Simulation operation**

When the simulation model is determined, a simulation program that is programmed in suitable computer language can be run on a computer. Because of the randomness of system, each simulation is just a realization of system performance. Therefore, the simulation should be run for many times to identify the statistical rule of the system. For different systems, different stopping criteria can be chosen.

(3) **Simulation results analysis**

The main method to analyze simulation results is the statistical analysis based on model behaviors. Generally, there are three kinds of model behaviors, i.e. the trajectory behavior, the structure behavior and the point behavior. The trajectory behavior means the change process of system states during simulation operation, which is often called as the system response. For example, the change of queue length as time advances in Fig. 9.1 is a trajectory behavior. Structural behavior is the system response due to the change of system structures. For example, if a pre-check process is added in Fig. 9.1, i.e. if the system structure is changed, the queue length as time advances will be completely different. This is a structural behavior. The point behavior corresponding to some comprehensive index of system response. For example, the mean queue length in Fig. 9.1 is a point behavior. For different simulation purposes, the model behaviors are different.

9.2.2 Simulation of Composite Lifeline Engineering System

The simulation technology of discrete event dynamic system provides a tool for studying the seismic response of composite lifeline system. The method can not only reflect the sudden change of system states in a seismic disaster field system but also can reflect the randomness of system properties and system responses. Actually, the discrete states and randomness are two basic characteristics of discrete event dynamic system. Meanwhile, when considering disaster prevention of a city, composite lifeline system is the major part of urban seismic disaster field. The seismic disaster response of a city is mainly determined by the interaction among different lifeline systems. Apparently, the interaction between different systems provides a realistic basis of system modeling, while the simulation of discrete event dynamic system provides a basic tool for system behavior analysis.

However, comparing with the simulation of classic discrete event dynamic system, the seismic disaster response simulation of composite lifeline system has its own particularity in the following aspects.

(1) The simulation of classic discrete event dynamic system generally considers simulation problem in the time domain, while the simulation of composite lifeline system must consider the disaster response process of several lifeline engineering systems during and after the earthquake. Therefore, the later must consider the spatial disaster distribution of lifeline systems besides the time domain simulation. In other words, it must be combined with the simulation of discrete event dynamic system in the time domain and network reliability analysis in the field domain for several branch lifeline systems.
(2) The simulation of classic discrete event dynamic system usually focuses on trajectory behaviors and point behaviors of the system. On the other side, as the target of the simulation of composite lifeline system is to find the weak link of the system for disaster prevention of a city, the simulation of composite lifeline system focuses more on system's structural behavior. In fact, the structural behavior is the basis of seismic disaster field control. Meanwhile, as the disaster index, an index to measure the system disaster capability is usually the function of system's behavior. So, the simulation of composite lifeline system should focus on the analysis of the structural behavior and the point behavior of the system.
(3) The descriptions of structural behavior in the simulation of discrete event dynamic system and composite lifeline system are similar. However, the change of system structure in the simulation of discrete event dynamic system results in the changes of queue rule or flow rule, while that in the simulation of composite lifeline system corresponds to the change of system spatial distribution.

9.2.3 Disaster Simulation Model of Composite Lifeline System

As described in Sect. 9.1, the constructing of a composite lifeline system from the perspective of secondary fire control can cover the main lifeline system and building block of a city. This section will supply a simulation model for such kind of systems.

In general, the systems related to the seismic secondary fire occurrence and firefighting include the urban building system, the firefighting system, the transportation system, the water supply system and the communication system. When considering the comprehensive disaster response of such systems, besides the secondary fire occurrence and fighting, the seismic damage of these systems themselves should be considered as well. Apparently, the prediction of the seismic damages of these systems belongs to the field domain simulation. For example, for building blocks, the seismic damage evaluation method such as fragility analysis method (Barbat et al. 1996; Singhal and Kiremidjian 1998) should be adopted to produce the distribution of the damages of a group of buildings. For transportation systems, the seismic connectivity evaluation between any two nodes should be implemented considering the damages of road and bridge. For water supply network, the nodal functional reliability should be given. For communication systems, the reliability of a major communication center should be evaluated. Meanwhile, for the simulation on time domain, a series of basic simulation parameters such as the receiving alarm time, setting out time, arriving time and fighting period and their probability model should be identified as well.

Considering the above background, two strategies, the coupling simulation strategy and the decoupling strategy, can be adopted for the disaster simulation of composite system.

In the coupling simulation, all spatial and time parameters are sampled simultaneously. This means that the seismic damages for different systems should be sampled randomly to construct a spatial simulation sample and the secondary firefighting process is sampled randomly to construct a time sample simultaneously. Then the two samples are hybrid with together to simulate the disaster response of the composite system. The process should be repeated for many times and then the results can be given by the statistical analysis.

Apparently, the simulation convergence process of the above strategy is very slow. In order to speed up the convergence rate, a decoupling simulation strategy can be adopted. In this strategy, the time domain simulation and field domain simulation are operated separately, which means that the time simulation of the system is frozen when simulating the spatial damages of building systems or the connectivity of a specified network. Then the simulation results of field domain are inserted into the simulation process of time domain. The decoupling simulation model is shown in Fig. 9.2, in which the secondary fire analysis model and the statistical parameters of firefighting can be found in Appendix C.

At the junction site of time domain and field domain, bulletin board technology and time delay technology can be used. Meanwhile, the colored Petri network introduced in the following section can be used as well to insert the simulation results of field domain in the simulation of time domain.

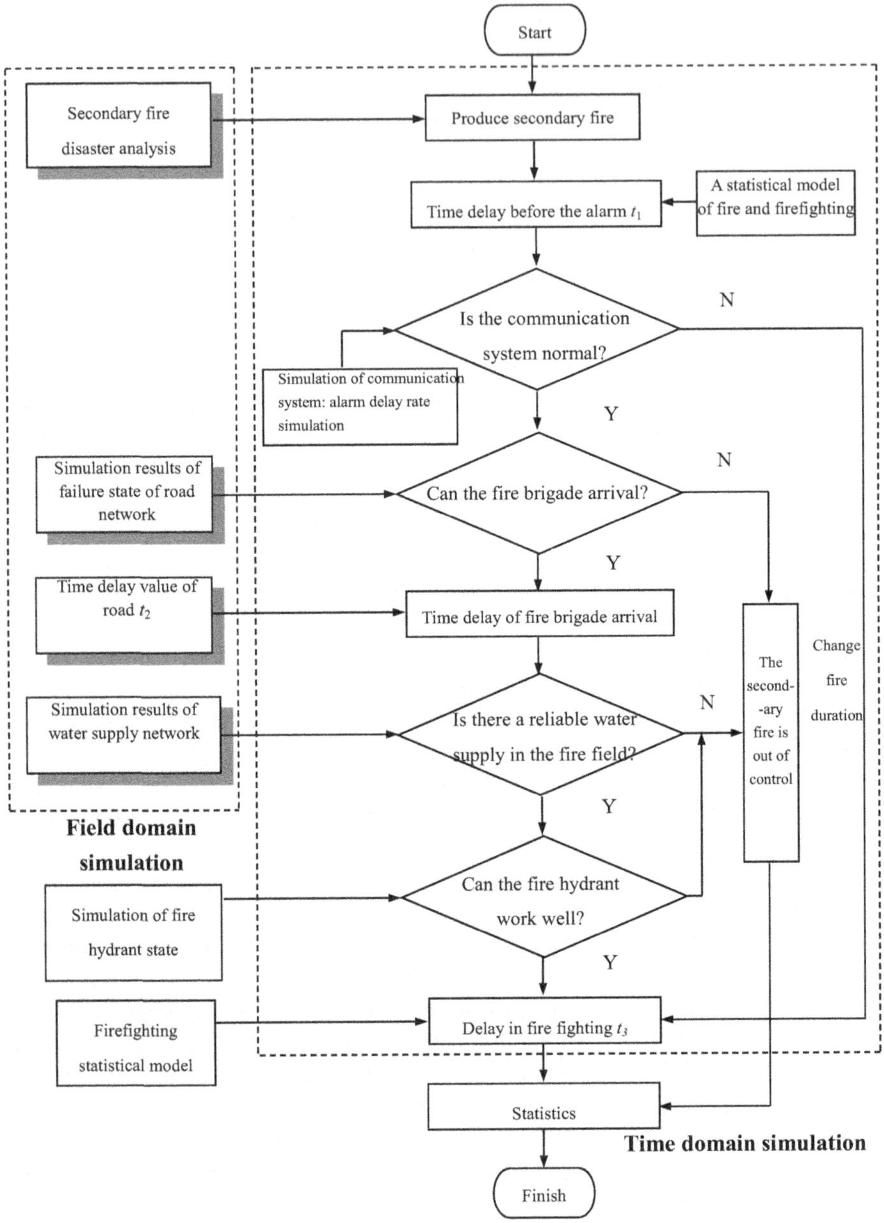

Fig. 9.2 Time-field domain simulation model of the composite lifeline system

9.2.4 Simulation Convergence Criteria and Simulation Statistics

Because of the randomness of system parameters, one simulation can only give one random sample of the system. In order to reflect the system function correctly, many samples should be generated and statistical analysis should be given to identify the correct system performance index. For the comprehensive disaster response of composite lifeline system, the mean fire lasting time, the total economic losses and the controlled secondary fire number can be chosen as the performance indices. Usually, two different simulation convergence criteria, mean convergence and covariance convergence can be adopted for a specified performance index. Let the general performance index be y, when n simulations are executed, the mean and covariance of y are, respectively:

$$\mu_{y_n} = \frac{1}{n}\sum_{i=1}^{n} y_i \tag{9.1}$$

$$\sigma_{y_n} = [\frac{1}{n}\sum_{i=1}^{n} (y_i - \mu_{y_n})^2]^{\frac{1}{2}} \tag{9.2}$$

Thus, the mean convergence criterion can be written as follows:

$$\sum_{i=n}^{m} \frac{|\mu_{y_{i+1}} - \mu_{y_i}|}{\mu_{y_i}} \leq \varepsilon_\mu \tag{9.3}$$

where m is a number which is larger than n and ε_μ is a prescribed error.

Similarly, the covariance convergence rule can be written as follows:

$$\sum_{i=n}^{m} \frac{|\sigma_{y_{i+1}} - \sigma_{y_i}|}{\sigma_{y_i}} \leq \varepsilon_\sigma \tag{9.4}$$

Obviously, the covariance convergence criteria are stricter than the mean convergence criteria and it takes more time to run simulations. Meanwhile, the number of simulations varies for different performance indices. Therefore, the identification of the convergence rule depends on the actual problem. When the only qualitative analysis is needed, reasonable simulation numbers can be given before running simulation.

For the simulation results which satisfy the simulation convergence criteria, statistical analysis can be carried out to give the point estimation, interval estimation and distribution frequency of the index.

9.3 Petri Net Model for Disaster Simulation of Composite Lifeline System

The model of the above simulation analysis is established based on physical flowchart or logical diagram. Furthermore, Petri net model (Peterson 1981), which is widely used in the simulation of discrete event dynamic system, provides a convenient tool for the time domain simulation of composite lifeline system, especially for the multi-resource and multi-entity systems.

9.3.1 Classic Petri Net

Petri net can simulate a discrete event dynamic system briefly and straightforward. Practically, it is suitable not only to delineate the change of system's structure and state but also to analyze the system dynamic characteristics.

As shown in Fig. 9.3, a Petri net is a binary graph determined by (**P**, **H**, **F**), where **P** is a set of position node p_i, **H** is a set of transition node h_i, $\boldsymbol{F}$ is a set of directed edge of $p \to h$ or $h \to p$. In Fig. 9.3, the position is represented by a circle and it usually represents the occurrence condition of an event. The transition is represented by "–" or "|" and it usually indicates the occurrence or termination of an event. The directed arc is represented by a line with an arrow and it indicates the relation between events. Net (**P**, **H**, **F**) indicates the system's structure and rule.

Petri net not only delineates the static relationship in system but also reflects dynamic characteristic of system by operation of the net. The system's dynamic

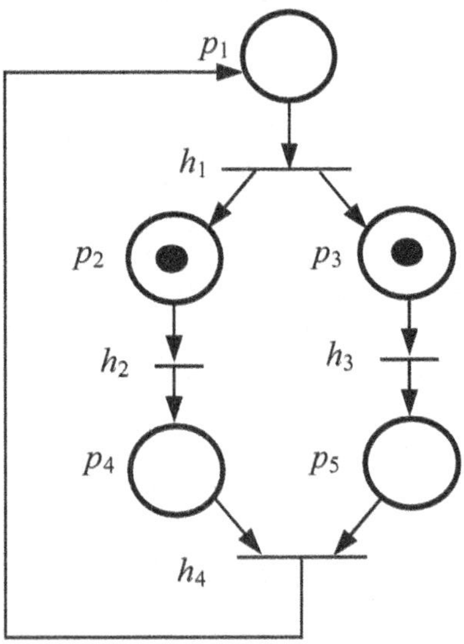

Fig. 9.3 Petri net diagram

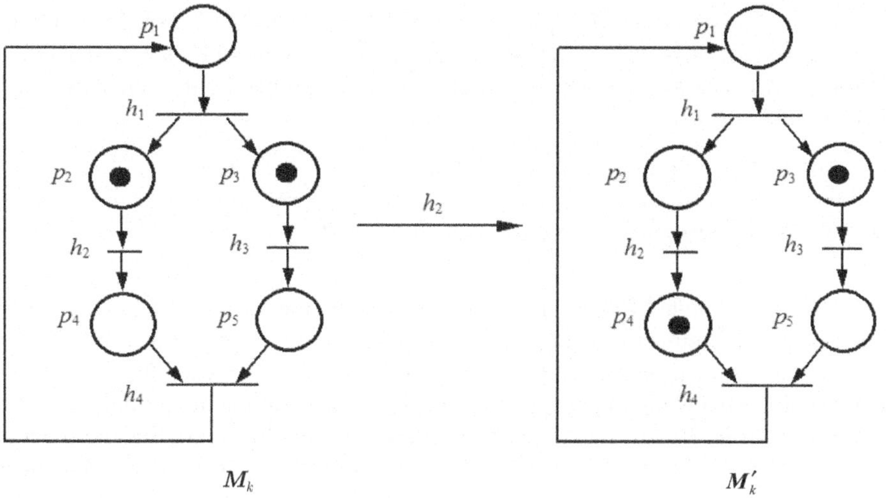

Fig. 9.4 The process of triggering transition

characteristic is realized by changing system's state, i.e. by triggering the transition. In fact, a transition owns an input location, $I(t_i)$, and an output location, $O(t_i)$. For example, for the transition h_2 in Fig. 9.3, its input location is p_2 and output location is p_4. The entities in each location are called marks (the black dot in the circle) and the number of marks in location p_i is represented by $M(p_i)$. Meanwhile, each arc can be given a weight to indicate the number of entities so that it can transfer. The weight of arc is represented by $w(p_i, h_j)$ or $w(h_j, p_i)$. In Fig. 9.3, the arc $p_2 \to h_2$ owns weight 1. Herein, the label is omitted when the weight equals to 1. When system state changes or a series of transitions are triggered, several marks, which equals to the weight of input arc, is removed from the input location to the corresponding output locations. For example, when transition h_2 is triggered in Fig. 9.3, the successive net is shown in Fig. 9.4.

Generally, $\mathbf{M(P)}$ is used to represent the marks in the net. After triggering transition, the marks in the net becomes $\mathbf{M'(P)}$. The relation between $\mathbf{M(P)}$ and $\mathbf{M'(P)}$ is

$$\mathbf{M'(P)} = \begin{cases} \mathbf{M(P)} - \mathbf{W(P, H)}, & \text{If } \mathbf{P} \in \mathbf{I}(t) \\ \mathbf{M(P)} + \mathbf{W(H, P)}, & \text{If } \mathbf{P} \in \mathbf{O}(t) \\ \mathbf{M(P)} + \mathbf{W(H, P)} - \mathbf{W(P, H)} & \text{If } \mathbf{P} \in \mathbf{I}(t) \cap \mathbf{O}(t) \\ \mathbf{M(P)} & \text{Others} \end{cases} \tag{9.5}$$

where $\mathbf{W(P, H)}$ is the weight matrix for input location arc and $\mathbf{W(H, P)}$ is the weight matrix for output location arc, $\mathbf{I}$ and $\mathbf{O}$ are the input and output, respectively.

Apparently, a series of transitions at different time points form a dynamic process of the system, which is represented by the moving path of $\mathbf{M(P)}$ as time advances.

Equation (9.5) describes this process from the mathematical perspective. Using this description, it is easy to design a simulation program and analyze the simulation results. The triggering rule of transition is determined by the actual physical background.

9.3.2 Non-Autonomous Colored Petri Net

With the enlargement of the scale of physical system, the scale of Petri net enlarges quickly, which limits the application of Petri net. In order to solve this problem, the marks in the same position can be distinguished by colors, i.e. each mark is related to one color. Meanwhile, the triggering function is set as a function of color, i.e. the triggering of transition is the triggering of a certain color. Such modified Petri net is called the colored Petri net (Jensen 1996). Moreover, when time is introduced into a colored and recorded when transition happens randomly in the time domain, the Petri net is called non-autonomous colored Petri net, which can be defined as follows:

$$NCPN = \left(\mathbf{P}, \mathbf{H}, \mathbf{W}^{+}, \mathbf{W}^{-}, \mathbf{C}, \mathbf{D}, \mathbf{M}\right) \tag{9.6}$$

where $\mathbf{P} = \{p_1, p_2, \ldots, p_n\}$ is the set of position and n is the number of positions,

$\mathbf{H} = \{h_1, h_2, \ldots, h_m\}$ is the set of transition and m is the number of transitions,

$\mathbf{W}^{+} = \mathbf{W}^{+}(\mathbf{P}, \mathbf{H})$ is the weighted arc matrix from colored position to colored transition and is called input function,

$\mathbf{W}^{-} = \mathbf{W}^{-}(\mathbf{H}, \mathbf{P})$ is the weighted arc matrix from colored transition to colored position and is called output function,

$\mathbf{C} = \left\{c(p_i), c\left(h_j\right)\right\}$ is the colored position and colored transition set, $c(p_i)$ is the colored set related to position p_i and $c(h_j)$ is the colored set related to transition h_j,

$\mathbf{D} = \{t_{h1}, t_{h2}, \ldots, t_{hm}\}$ is the random time set, i.e. the random time for each transition in the net. Each random time represents the time from the beginning to the end of triggering.

$\mathbf{M}$ is the colored mark, i.e. the system state function.

When the transition h_j for color c_k is triggered, it is recorded as h_j/c_k and leads to following two operations:

(1) All input locations p_i of h_j subtract the number of marks $\mathbf{W}^{+}\left(p_i, h_j/c_k\right)$;
(2) All output locations p_i of h_j add the number of marks $\mathbf{W}^{-}\left(h_j/c_k, p_i\right)$.

Let $\mathbf{M}_{k+1}$ represent the mark after h_j for color c_k is triggered. Then the relationship between $\mathbf{M}_{k+1}$ and $\mathbf{M}_{k}$, the mark before h_j is triggered, is

$$\mathbf{M}_{k+1}(p_i) = \mathbf{M}_{k}(p_i) + \mathbf{W}^{-}\left(h_j/c_k, p_i\right) - \mathbf{W}^{+}\left(p_i, h_j/c_k\right) \tag{9.7}$$

When a transition is triggered for a certain color, all colored marks related to the arc of this transition are changed.

Non-autonomous colored Petri net is driven by triggering a series of transitions. When transition h_1 is triggered for color set c_{l1}, it is recorded as h_1/c_{l1} and results in that mark $\mathbf{M}_1$ becomes $\mathbf{M}_2$ after random time $\mathbf{D}_{l1}$. Subsequently, when transition h_2 is triggered for color set c_{l2}, it is recorded as h_2/c_{l2} and results in that mark $\mathbf{M}_2$ becomes $\mathbf{M}_3$ after random time $\mathbf{D}_{l1}$. At last, after a series of triggers

$$S = (h_1(\mathbf{D}_{l1})/c_{l1}, h_2(\mathbf{D}_{l2})/c_{l2}, \ldots, h_k(\mathbf{D}_{lk})/c_{lk}) \tag{9.8}$$

the mark becomes

$$\mathbf{M}_{k+1}(\mathbf{P}) = \mathbf{M}_1(\mathbf{P}) + \sum_{j=1}^{k}\left[\mathbf{W}^-\left(h_j(\mathbf{D}_{lj})/c_{lj}, \mathbf{P}\right) - \mathbf{W}^+\left(\mathbf{P}, h_j(\mathbf{D}_{lj})/c_{lj}\right)\right] \tag{9.9}$$

Apparently, Eqs. (9.7)–(9.9) construct a set of equations which actually give a mathematical description of discrete event dynamic system. The series $\mathbf{M}_1$, $\mathbf{M}_2$,..., $\mathbf{M}_k$,... constructs the system tracks of a simulation.

9.3.3 *Seismic Disaster Simulation of Composite Lifeline System*

Based on the above non-autonomous colored Petri net, a time domain simulation model of composite lifeline system, as illustrated in Fig. 9.5, can be established. The elements in the position set **P** and the transition set **H** are shown in Tables 9.2 and 9.3, respectively. For the occurrence and fighting process of secondary fire, the state change is represented by the appearance and disappearance of colored element in the positions and transitions, i.e. the flow of the marks. The marks that represent the fire states and firefighter's states move from one transition to another transition means the state change of the system.

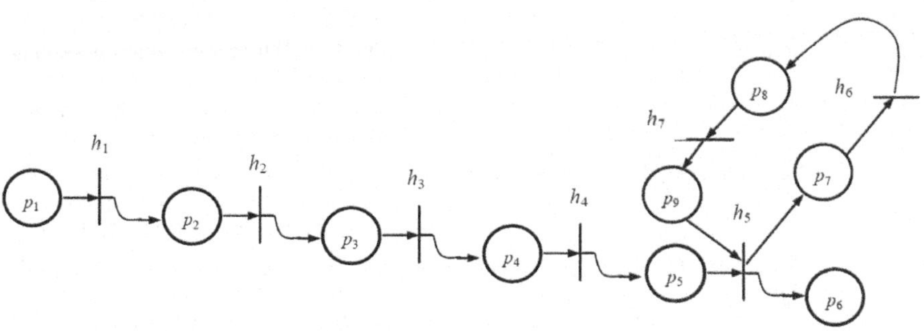

Fig. 9.5 Non-autonomous colored Petri net model for the simulation of compound lifeline system

Table 9.2 Element of position set and its corresponding colored element

Position	Position element	Colored element
p_1	The occurrence of secondary fire	Fire NO., Occurrence time, Occurrence location
p_2	Command center knows the occurrence of fire	Fire NO., Occurrence time, Occurrence location
p_3	Judge fire category	Fire NO., Fire category
p_4	Responsible firefighting team leaves for the fire location	Firefighting team, Firefighting trucks, Firefighting team address
p_5	Responsible firefighting team arrives at the fire location	Fire NO., Firefighting team
p_6	Secondary fire is put off	Fire NO., Firefighting team
p_7	Knowing the fire details	Fire NO., Fire details
p_8	Command center knows the fire details	Fire NO., Fire information on the road, Fire details, Fire information of putting off
p_9	Assistant firefighting team arrives at the fire location	Fire NO., Firefighting team

Table 9.3 Element of transition set and its corresponding colored element

Transition	Transition element	Colored element
h_1	Receive fire alarm	Fire NO., Occurrence time, Occurrence location, Weather, Wind, time of receiving fire alarm
h_2	Judge fire category	Fire NO., Occurrence time, Occurrence location, Weather, Wind, time of judging fire alarm
h_3	Ask responsible firefighting teams to fight	Fire NO., Fire category, Firefighting team, Firefighting team address, time of manoeuvring responsible firefighting team
h_4	Go to fire location	Fire NO., Fire category, Firefighting team, firefighting team address, Time of leaving, Arrival time, Time on the road
h_5	Fight the fire	Fire NO., Whether enough water can be supplied at the fire location, Beginning and end time of putting off fire, Random time of putting off fire
h_6	Report fire details to command center	Fire NO., Fire information on the road, Fire information after arriving fire location, time of reporting fire details
h_7	Ask assistant firefighting teams to help to fight fire	Fire NO., Fire information on the road, Fire information after arriving fire location, Firefighting team, Firefighting team address, time of manoeuvring assistant firefighting team

Obviously, the seismic damage influence of all systems can be inserted into simulation by colored elements, which represent the fire occurrence and alarm times, whether firefighting trucks can arrive at the fire locations, the time that firefighting trucks run on the road and whether enough water can be supplied, etc. Then a time-field domain simulation model for the seismic disaster simulation of a city can be constructed.

Considering the spatial distribution of fire, the damage of building and lifeline engineering systems, billboard technology and queue theory is introduced to assemble the problem of single-resource multi-entity simulation into the problem of multi-resource multi-entity simulation. That is, the firefighting teams are commanded by a command center (billboard) to establish several parallel Petri net simulation models. In this way, the target city can be divided into several simulation districts and the occurrence of secondary fire of the city districts is simulated according to the building seismic damages estimation and seismic secondary fire risk model as the time (simulation clock) advances. The secondary fire is recorded at billboard and the responsible firefighting team which is close to the fire location is manoeuvred and corresponding Petri net simulation model is triggered. When any of the following cases, fire alarm failure, traffic cutoff and no water or fire hydrant failure, happens, the secondary fire will get out of control. Then, a random time duration with uniform distribution can be considered to simulate the fire lasting time. Otherwise, the execution time of Petri net is the fire lasting time. When the responsible firefighting team has not put off the fire, a new fire occurs for which the current firefighting team is responsible, the neighbor firefighting team is manoeuvred. If no firefighting team is available for the firefighting, the new fire is inserted into the system queue and recorded the last time. When a fire fighting team is available, a fire event is simulated according to first-in-first-out rule. Meanwhile, when a fire team needs help, billboard is used to maneuver assistant fire fighting team. The above process constructs a simulation process of multi-resource-multi-entity. During the process, the seismic connectivity reliability of each lifeline system is calculated and is used as the colored elements of Petri net.

9.4 Case Study on Seismic Disaster Simulation

Taking a typical city area, shown in Fig. 9.6, as an example, the seismic responses of lifeline engineering systems are studied and the secondary firefighting processes are simulated. The area is 113 km^2 and is divided into 35 simulation zones according to the main road network. For each engineering system, the earthquake damages of elements are predicted and the network reliability analysis is carried out.

(1) **Water supply system**

According to the function reliability analysis results, 156 nodes are in seriously damage state in the water supply system under the seismic intensity of IX.

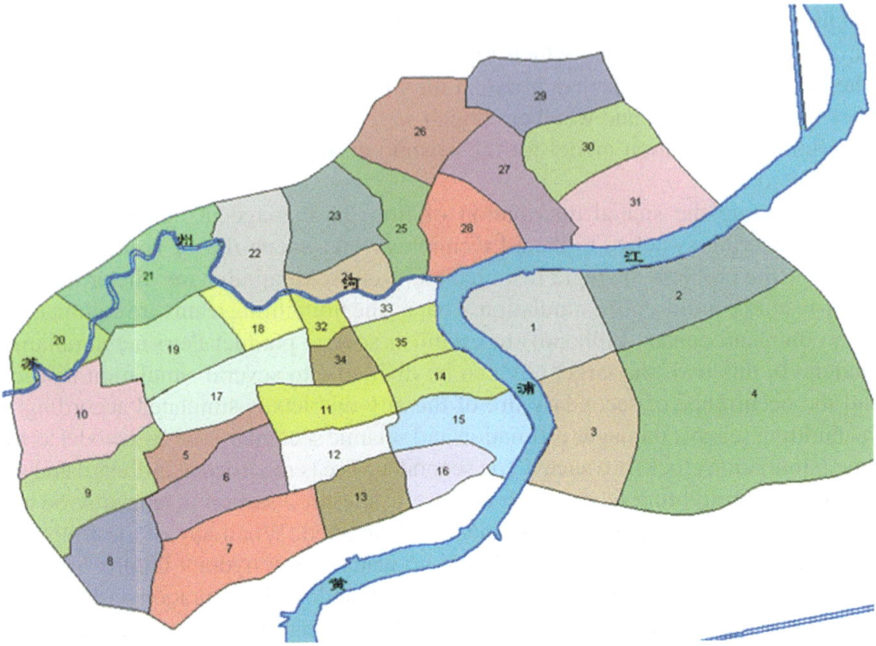

Fig. 9.6 Locations and numbers of simulation zones

The influence of water supply system on the secondary firefighting is simulated by the following ways. Let zone i have n nodes. If two thirds of nodes are in the states of serious unreliable or cutoff, the secondary fire in zone i will be out of control. Otherwise, the water supply in zone i is available for firefighting. For zones which are out of control, the fire duration is assumed to be a random variable uniformly distributed within a certain period (such as 4~30 h).

(2) **Transportation system**

According to the connectivity reliability analysis, 16.43% roads of the main roads in the area are in seriously unreliable states under seismic intensity IX. Supposing there are n nodes around zone i, when the connectivity reliability between the nodes and the fire brigades is less than 0.7, it is assumed that the fire brigades are unable to reach zone i. Otherwise, the fire brigades are assumed to be able to reach zone i.

(3) **Firefighting system**

Considering the coverage rate of fire hydrants, it is predicted that 80% fire hydrants can work after earthquakes, i.e. more than 20% fires will be out of control because of the failure of hydrants.

(4) **Communication system**

Both damages in core buildings and communication equipment will influence the post-earthquake function of communication systems. According to the analysis, the

Table 9.4 Secondary fire simulation of a typical zone

Secondary fire No	Simulation zone No	Time of fire	Location of fire	Time from fire to alarm (min)	Time from alarm to arrival (min)	Time from arrival to firefighting (min)	Duration (min)	Loss (ten thousand Yuan (¥))
2–2	2	12:16	II	3	10	16	30	5.58

main communication buildings in the area are in the states of lower damage due to the reasonable seismic design. Therefore, only the damage of communication equipment is considered. Herein, the function rate of communication system is predicted to be 80%, i.e. the 20% fires cannot be captured by alarm in time due to the failure of communication equipment. For the timely alarm, the alarm rate is simulated by a random variable uniformly distributed in [0, 1].

(5) **Building system**

In addition to the above lifeline engineering systems, the damage prediction of buildings in the area is also required to calculate the occurrence rate of secondary fire. Based on the prediction, the occurrence rate of post-earthquake secondary fire in all zones are calculated and then used in simulation.

Using the above results, the seismic disaster field in the area is simulated. The simulations are repeated for 500 times independently to reach the steady convergence condition. Table 9.4 shows typical simulation results in a zone. And Table 9.5 shows the whole simulation results. In the table, "■" and "▲" mean that the secondary fires are out of control because of the failures of water supply system or transportation system, respectively. "√" represents that the corresponding systems have normal function. In addition, the fire loss is calculated by the empirical formula proposed by Li and Song (1996a, b).

It is shown that there are 13 simulation zones suffer from the failure of water supply or transportation systems. In these zones, 195 secondary fires occur in three days after the earthquake with the mean loss of 3.08 million Yuan for each fire. For the other zones, though 192 secondary fires occur as well, the mean duration of each fire is reduced obviously because the water supply and transportation systems function keep in normal state. As a result, the mean loss is only 0.65 million Yuan for each fire, which is only 21% of the former. Therefore, whether lifeline systems work well after earthquakes are critical for the earthquake resistance and post-earthquake relief. Before earthquakes, to identify and retrofit the weak parts of a city according to the seismic reliability analysis of lifeline systems and the seismic response simulation of composite lifeline engineering systems is vital for improving the seismic resilience of a city.

Table 9.5 Seismic response simulation results of composite lifeline engineering systems

Simulation zone No	Water supply system	Transportation system	Occurrence times	Average duration/sec	Function rate of communication system/%	Function rate of firefighting system/%	Firefighting success rate/%	Loss/ (ten thousand Yuan (¥))
1	√	√	16	2018	87.5	81.8	69.4	375.35
2	√	√	19	2231	84.8	86.5	71.3	414.97
3	√	√	17	2725	85.9	87.2	73.1	506.85
4	√	√	57	3943	77.6	81.4	59.2	733.40
5	■	√	9	18,680	—	—	0	3474.51
6	√	√	20	3052	82.8	84.6	67.4	567.68
7	√	√	30	3405	84.7	76.2	61.9	633.32
8	√	√	10	1793	87.2	84.9	72.1	333.50
9	■	√	20	29,202	—	—	0	5431.24
10	√	√	25	2769	80.0	83.7	63.7	515.03
11	■	√	11	19,486	—	—	0	3624.46
12	√	▲	9	18,925	—	—	0	3520.18
13	√	▲	10	20,256	—	—	0	3767.62
14	√	√	10	1908	88.9	78.9	67.8	354.89
15	√	▲	16	24,327	—	—	0	4523.65
16	√	▲	9	18,353	—	—	0	3413.76
17	■	√	10	19,762	—	—	0	3675.77
18	√	√	5	1018	87.7	85.1	72.8	189.35
19	■	▲	11	19,805	—	—	0	3683.71
20	√	▲	23	34,178	—	—	0	6357.16

(continued)

Table 9.5 (continued)

Simulation zone No	Water supply system	Transportation system	Occurrence times	Average duration/sec	Function rate of communication system/%	Function rate of firefighting system/%	Firefighting success rate/%	Loss/ (ten thousand Yuan (¥))
21	√	▲	44	65,643	—	—	0	12,210.10
22	√	√	23	2435	78.5	82.7	61.5	452.58
23	√	▲	25	34,596	—	—	0	6433.56
24	√	√	12	1415	87.4	85.6	73.0	263.19
25	√	√	11	1686	85.9	82.7	68.6	313.60
26	√	√	17	2327	76.5	87.8	64.3	432.82
27	√	√	16	2253	86.6	82.3	68.9	419.06
28	√	√	14	1728	85.3	87.1	72.5	321.41
29	√	√	22	2659	86.2	79.5	65.7	494.57
30	√	√	20	2582	82.7	85.1	67.8	480.25
31	√	√	25	2603	86.6	76.8	63.4	484.16
32	√	√	4	845	87.9	86.7	74.6	157.17
33	√	√	5	896	88.3	87.1	75.0	166.65
34	√	√	5	913	84.5	85.6	73.7	169.82
35	√	√	8	1438	78.6	87.5	67.2	267.47
Total			387				68.1	72,635.36

9.5 Urban Earthquake Disaster Field Control

It has been mentioned in the Sect. 9.1 that the goal of seismic disaster response simulation of composite lifeline system is to investigate the comprehensive disaster resilience capability of lifeline engineering systems. Moreover, the simulation can be used to determine the strategies to improve the aseismic capability of lifeline engineering systems. In other words, the strategies to improve disaster resilience of a modern city can be obtained with help of the seismic disaster simulation of composite lifeline system (Li and Song, 1996b).

9.5.1 System Control Based on Structural Behavior

System structural behavior means that its trajectory and point behaviors change with the system structure. There are two methods that could be used for system's structure behavior analysis i.e., the sub-system analysis and the sub-structure analysis. The two methods at different levels construct a complete hierarchical control model and the corresponding algorithm as shown in Fig. 9.6. Similar to the feedback control of discrete event dynamic system, the functional feedback can be used for system control. At the sub-system level, the prior retrofit sub-system is determined by investigating the change of compound system's behavior. For the prior retrofit sub-system, reliability analysis and optimization technology are used to determine the detailed retrofit measures. Apparently, during this process, the basic feedback mechanism is to investigate the disaster response of system by changing system structure or sub-system structure. According to the system response, the control rule can be determined. Such an excitation-control circulation constructs a complete system control process (Fig. 9.7).

Take the simulation model of the composite system described in Sect. 9.2 as a basic tool and taking the secondary fire disaster loss as an evaluation index. Figure 9.8 shows two examples for two cities of China (Li 1997). These case studies show that the control strategy and process may be significantly different for different cities.

The control benefit index of disaster field may be defined as follows:

$$\lambda = \frac{L_b - L_a}{C} \tag{9.10}$$

where L_b and L_a are the seismic disaster loss before and after system control, respectively, C is the investment for disaster control.

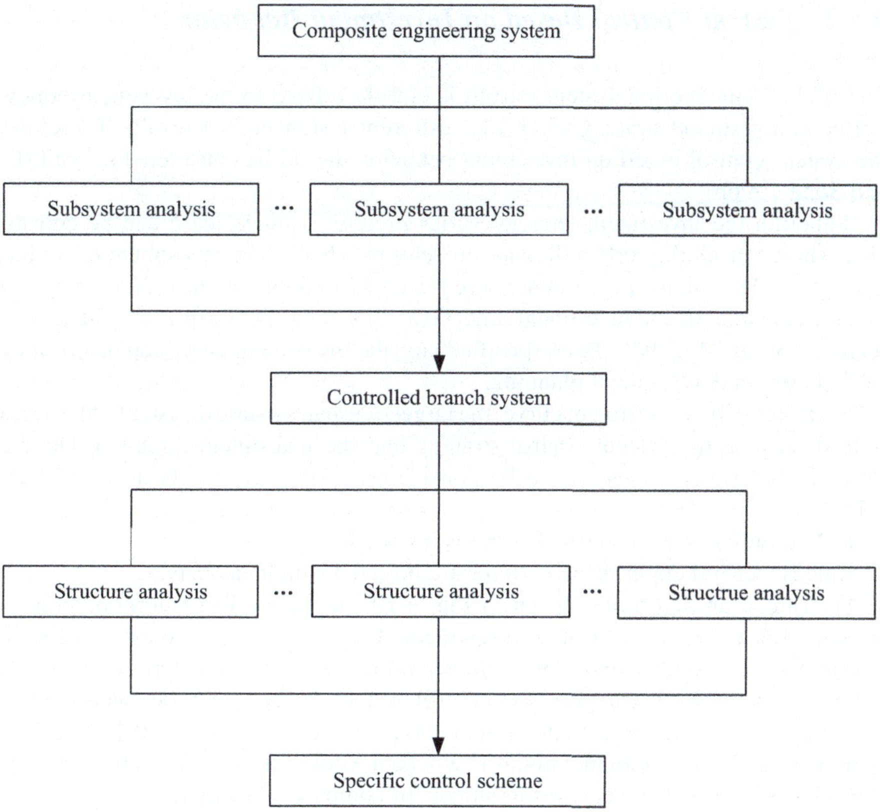

Fig. 9.7 Hierarchical control model

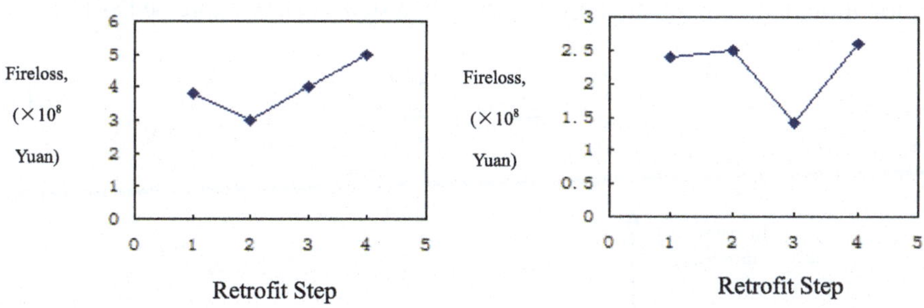

(a) Analysis result for Kaifeng City (b) Analysis result for Puyang City

(Horizontal axis: 1. Retrofitting water distribution network, 2. Retrofitting transportation system, 3. Retrofitting firefighting system, 4. Retrofitting communication system

Fig. 9.8 Case studies of simulation control of urban seismic disaster field

9.5.2 System Control Based on Investment Behavior

The lifeline engineering system retrofit is closely related to the investment policy. Different investment strategy will lead to different system control results. Therefore, the system control based on investment behavior should be considered as well (Li and Song 1996b).

Although the investment strategy varies greatly, a simple case can be considered when considering urban disaster prevention. That is, the investment curve has been given. Herein, the investment curve is a curve to illustrate the relation between investment strength and investment time. Generally, it can be illustrated by an investment histogram (Fig. 9.9). For a specified city, the investment curve can be given by resilient urban development planning.

For a prescribed investment curve, the target of urban seismic disaster field control is to determine the system control strategy and the investment direction. Herein, different investment rules should be considered. There are two basic investment rules.

Rule I, only one engineering system is invested;

Rule II, several engineering systems are invested simultaneously.

For rule I, an algorithm shown in Fig. 9.10 can be used to determine control strategy. Let C_c be the cost of an investment. Then C_c can be used as a constraint to optimize each sub-system of the composite lifeline system to determine available control strategy. In the process, the control benefit index λ_i can be calculated by operating disaster response simulation of composite lifeline system one by one. The optimization strategy with maximum λ_i will be the optimum system control strategy. This algorithm is called the investment benefit comparison method.

For investment rule II, the hierarchic control theory of large system can be used to design a control algorithm (Li 1993). Assume there are m sub-systems. Then the distribution of an investment for different sub-systems can be expressed as follows:

$$c_l = \sum_{j=1}^{m} c_{jl} = \sum_{j=1}^{m} a_j c_l \tag{9.11}$$

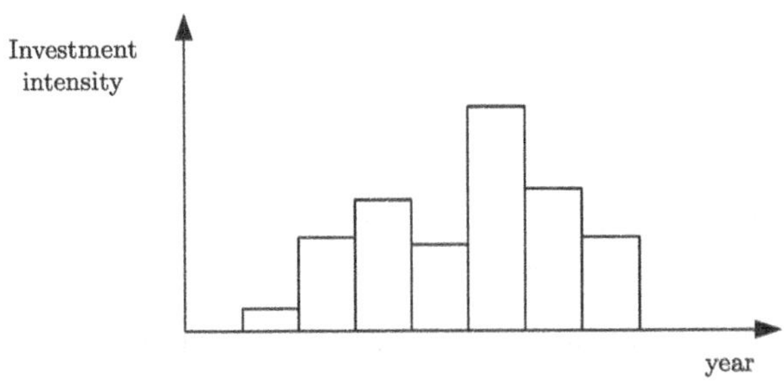

Fig. 9.9 Phased investment histogram

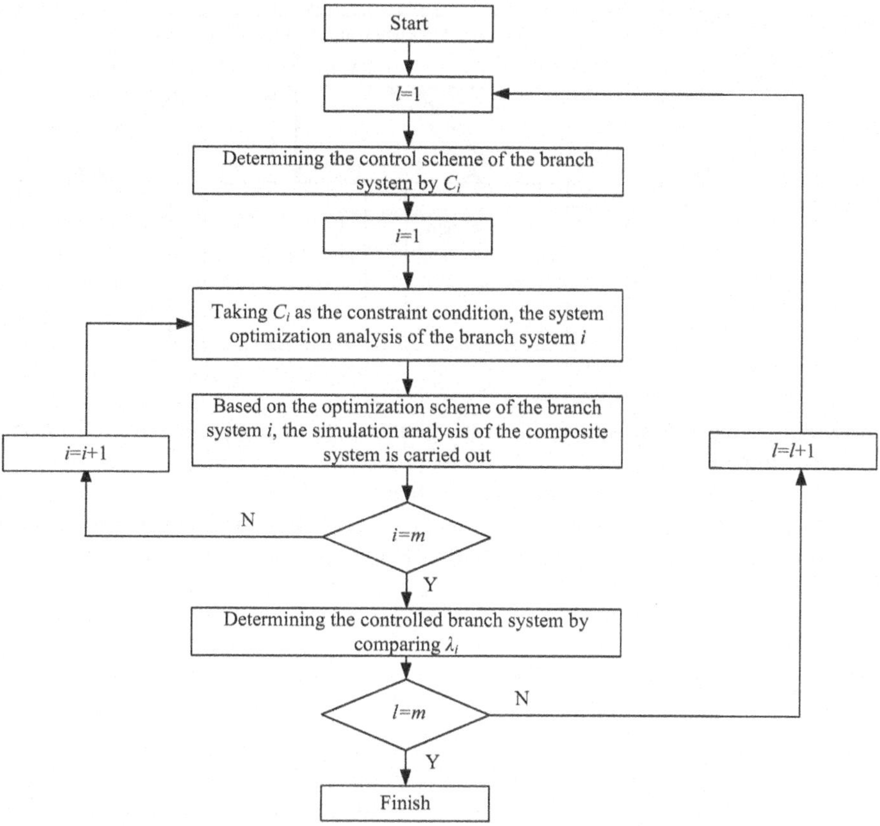

Fig. 9.10 Investment benefit comparison method-based on the investment rule I

Initializing a_j and optimizing the sub-system with c_{jl} as a constraint, an available optimization control strategy can be determined. Based on the optimization control strategy, the disaster response of compound lifeline system is simulated to calculated the control benefit index λ_j. Then, a new a_j is given as follows:

$$a_j = \frac{\lambda_j}{\sum_{i=1}^{m} \lambda_i} \tag{9.12}$$

The investment is redistributed and the above steps are repeated until the best control benefit is achieved. The flowchart of this algorithm is shown in Fig. 9.10.

There are two levels of control in Fig. 9.11. The first level charges to balance and distribute the total investment, and the second level determines the control strategy of the branch lifeline systems. Actually, on the second level, during the simulation of composite system after obtaining the optimization strategy of sub-systems, all the other sub-systems are in the phase of previous investment being finished. While, on the first level, all the sub-systems are in the optimization states after distributing all

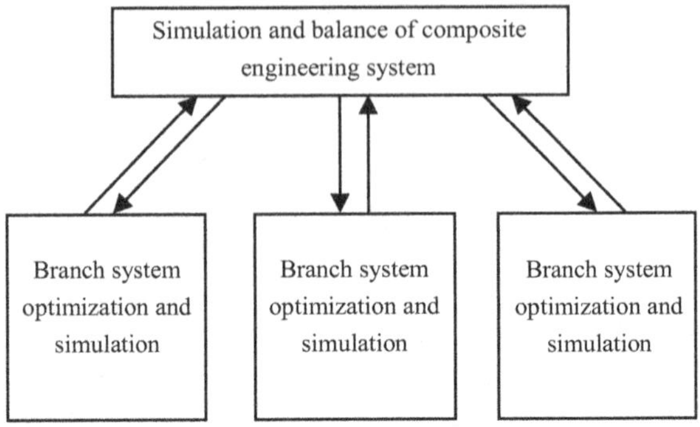

Fig. 9.11 Hierarchic control method based on the investment rule II

current investment. In this way, the investment benefit of comprehensive retrofitting strategy can be given.

Either investment benefit comparison method or hierarchic control method is control algorithms based on system functional feedback. These kinds of control algorithms are called the system simulation-based control.

9.5.3 Case Study

Taking Kaifeng city, which is located in the central part of China, as an example, the control of urban earthquake disaster field is studied (Li and Song 1996b). The area of the city is 60 km^2, which is divided into 26 simulation zones. As the foundation of system control analysis, the following studies are completed in advance, including the seismic reliability analysis of water supply and transportation systems, the seismic vulnerability prediction of communication system, the seismic capacity analysis of firefighting system, the seismic damage prediction of the building blocks and the fire hazard analysis of the city.

Then, the seismic response of the composite lifeline engineering system is studied considering post-earthquake fire disaster and taking the results as the basic state of the disaster field. On this basis, the control strategy about lifeline engineering system is studied using the system structural behavior-based control strategies.

(1) **First stage control**
(a) Water supply system control

Based on the seismic reliability analysis of water supply system in the city, the secondary fire in seven simulation zones are out of control due to the failure of water supply system in the basic disaster field state. According to the principles of structural behavior-based control, only the water supply system is retrofitted to guarantee the water supply of the above seven simulation zones, and the other lifeline engineering systems are still kept in the original states (without retrofitting). On this basis, the

Table 9.6 Disaster state by controlling water supply system

Number of secondary fire/times	Average duration/min	Firefight success rate/%	Loss/ten thousand Yuan (¥)
176	1166	60.9	38,279

Table 9.7 Disaster state by controlling transportation system

Number of secondary fire/times	Average duration/min	Firefight success rate/%	Loss/ten thousand Yuan (¥)
174	984	67.7	31,917

new disaster state can be given by disaster field simulation again and the results are given in Table 9.6.

(b) Transportation system control

Based on the seismic connectivity reliability analysis of road network of the city, the secondary fire in the simulation zone 1 is out of control due to the failure of transportation system. Herein, only transportation system is retrofitted so that the fire brigades can reach the simulation zone 1 after the earthquake. Meantime, the other lifeline engineering systems are still kept in the basic state of the disaster field (without retrofitting). Then, the disaster state can be given by simulating the disaster field again and Table 9.7 shows the results.

(c) Firefighting system control

In the basic disaster field state, only 80% fire hydrants work well because of hydrants failure. Similarly, retrofit the firefighting system and make that 100% fire hydrants work well, while the other lifeline engineering systems are still in the basic disaster field states. The disaster state can be given by simulating the disaster field again and Table 9.8 is the results.

(d) Communication system control

Based on the damage prediction results of core buildings and equipment in the communication system, the alarm success rate of secondary fire is 80%. Therefore, the communication system is retrofitted to improve the alarm success rate up to 100%, and the other systems remain the basic disaster field states. The disaster field simulation results are shown in Table 9.9.

Table 9.8 Disaster state by controlling firefighting system

Number of secondary fire/times	Average duration/min	Firefight success rate/%	Loss/ten thousand Yuan (¥)
176	1257	57.8	41,065

Table 9.9 Disaster state by controlling communication system

Number of secondary fire/times	Average duration/min	Firefight success rate/%	Loss/ten thousand Yuan (¥)
175	1525	47.2	49,678

Table 9.10 Disaster state by controlling building engineering system

Number of secondary fire/times	Average duration/min	Firefight success rate/%	Loss/ten thousand Yuan (¥)
129	1559	47.0	37,470

(5) Building blocks

Under the seismic intensity of IX, 92.8% buildings are in the state of moderate or serious damage. After retrofitting old buildings and construction of new buildings, the damage area reduces by one third after earthquake. The disaster field simulation results are shown in Table 9.10.

By comparing the above control strategies, it is found that the transportation system should be retrofitted first to achieve the most benefit.

After the controlled branch system is determined, the corresponding engineering strategies can be given by seismic reliability optimization of the single lifeline engineering system. According to the analysis, the retrofitted roads are selected. By retrofitting these roads, the connection between the simulation zone 1 and firefighting brigades is guaranteed.

(2) Second stage control

After the control in the first stage is accomplished, the resulting disaster field state can be seen as an initial state to conduct the next stage system control analysis. Similar to the above procedure, the disaster state by controlling different systems can be obtained as listed in Table 9.11. Note that the influence of transportation system is not considered in the second stage because it has been retrofitted in the first stage and will work well in the second stage.

Table 9.11 State comparisons by controlling different systems (the second stage control)

Control system	Number of secondary fire/times	Average duration/min	Firefight success rate/%	Loss/ten thousand Yuan (¥)
Communication system	178	945	67.7	31,337
Water supply system	175	620	81.1	20,192
Firefighting system	176	566	82.9	18,593
Building engineering system	134	995	67.3	24,742

Table 9.12 State comparisons by controlling different systems (the third stage control)

Control system	Number of secondary fire/times	Average duration/min	Firefight success rate/%	Loss/ten thousand Yuan (¥)
Communication system	178	515	82.9	17,054
Water supply system	177	94	100	3100
Building engineering system	132	578	82.4	14,254

Based on the result, the firefighting system is selected as the controlled system in this stage. In order to do so, 550 fire hydrants are added and the corresponding daily maintenance system is established.

(3) Third stage control

After the retrofitted strategy in the second stage is accomplished, the last state can be taken as the initial state of the next round. Table 9.12 shows the simulation results by controlling different systems. Herein, the firefighting and transportation systems are considered to work well in the stage.

As shown in Table 9.12, the water supply system should be the controlled system in this stage. In principle, the function reliability-based network optimization methods can be used to optimize the water supply system. However, the retrofit strategies can also be directly determined based on the comparisons among some potential strategies. Herein, the latter is selected. Based on the seismic reliability analysis of water supply system and the comparative analysis of retrofit schemes, the control goal in this stage is realized.

(4) Fourth stage control

Theoretically, the control system in the fourth stage can be determined by similar procedure as above. However, noticing that the control system in this stage only includes communication system and building system, comparing analysis may be more suitable in the stage. Apparently, reducing the occurrence rate of secondary fire by improving the seismic capacity of building system requires huge retrofit cost and long time. Therefore, the control system in this stage should be the communication system. By enhancing the core buildings and equipment, the communication system is improved and the final disaster field state is shown in Table 9.13.

Based on the control strategies in different stages, the phased retrofit plans of Kaifeng city are summarized in Table 9.14.

The above case study indicated that simulation-based system control provides an efficient tool for improving the seismic resilience of a city and it is available in practice. In this example, by taking system control, the firefighting success rate increases from 43.6% to 100%, and the loss of secondary fire reduces from 507.25 million to 10.22 million. Moreover, by retrofitting the water supply system, the transportation system and the communication system, the reliability of these lifeline engineering

Table 9.13 Final disaster field state (the fourth stage control)

Control system	Number of secondary fire/times	Average duration/min	Firefight success rate/%	Loss/ten thousand Yuan (¥)
Communication system	176	31	100	1022

Table 9.14 Seismic retrofit plan of Kaifeng city

No	System	Cost/ten thousand Yuan (¥)	Reduced loss/ten thousand Yuan (¥)	Benefit λ	Retrofit strategies
1	Transportation system	3200	18,808	5.9	Broadening Xihuan Road, Beihuan Road, Beimen Street and West Tieta Street, and retrofitting the buildings along these roads to improve the traffic capacity of fire trucks
2	Firefighting system	200	13,324	66.6	Set one fire hydrant every 120 m along the main water supply pipelines; Establish the maintenance management system of fire hydrant
3	Water supply system	272	15,493	57.0	Retrofit the diameter of 27,200 m long pipelines which may be seriously damaged in the earthquake
4	Communication system	100	2078	20.8	Enhance the equipment supporting structures and the core buildings of communication system

systems improves obviously, and therefore the aseismic resilience of the city can be improved. All these results prove the value of the control theory of urban earthquake disaster field.

References

Barbat, A. H., Moya, F. Y., & Canas, J. A. (1996). Damage scenarios simulation for seismic risk assessment in urban zones. *Earthquake Spectra, 12*(3), 371–394.

Han, X., Li, J., & Shen, Z. Y. (2000). Non-autonomous coloured Petri net-based methodology for the dispatching process. *Fire Safety Journal, 35,* 299–325.

Jensen, K. (1996). *Colored petri nets.* Berlin: Springer-Verlag, Berlin Heidelberg.

Kameda, H. (1994). Multi-phase evaluation of lifeline system performance under earthquake environment. In *2nd China-Japan-US trilateral symposium on lifeline earthquake engineering.* Xian.

Li, J. (1991). The system method of the earthquake disaster field control. *Journal of Catastrophology, 9*(1), 84–88 (in Chinese).

Li, J. (1992). *Earthquake disaster prediction and disaster prevention planning.* Henan Science and Technology Press. (in Chinese).

Li, J. (1993). Research on system control theory of earthquake disaster field. Project summary report of national natural science foundation of China. (in Chinese).

Li, J. (1997). Disaster response analysis and system control of composite engineering system. *Journal of Natural Disaster*, 6(3): 1–9. (in Chinese).

Li, J., & Song, J. X. (1996a). Study on the simulation and simulation-based control for earthquake disaster field part 1: The simulation of earthquake disaster field. *Earthquake Engineering and Engineering Vibration, 16*(2), 46–54. (in Chinese).

Li, J., & Song, J. X. (1996b). Study on the simulation and simulation-based control for earthquake disaster field part 2: Simulation-based control of earthquake disaster field. *Earthquake Engineering and Engineering Vibration, 16*(4), 21–29. (in Chinese).

Nozhati, S., Ellingwood, B. R., & Mahmoud, H. (2019). Understanding community resilience from a PRA perspective using binary decision diagrams. *Risk Analysis, 39*(4), 2127–2142.

Ouyang, M., & Wang, Z. (2015). Resilience assessment of interdependent infrastructure systems: With a focus on joint restoration modeling and analysis. *Reliability Engineering & System Safety, Special Issue on Resilience Engineering, 141,* 74–82.

Peterson, J. (1981). *Petrinet theory and the modelling of systems.* PreenticeHall: Englewood.

Scawthorn, C. (1992). Lifeline interaction and post-earthquake functionality. In *5th U.S.-Japan workshop on earthquake disaster prevention for lifeline systems.* Japan: Tsukuba Science City.

Singhal, A., & Kiremidjian, A. S. (1998). Bayesian updating of fragilities with application to rc frames. *Journal of Structural Engineering, 124*(8), 922–929.

Shinozuka, M., & Tanaka, S. (1996). Effects of lifeline interaction under seismic conditions. *11th World Conference on Earthquake Engineering.*

Appendix A
Boolean Algebra Basic

A.1 Boolean Variable

Boolean variable takes only two values, 0 and 1. Usually, 0 represents that the element fails and 1 represents that the element operates.

A.2 Basic Operation

There are three basic operations for Boolean variables. Any complex operation can be expressed by combining these three operations.

A.2.1 Union Operation

Union operation is usually called the logic sum. The union operation of two Boolean variables x and y can be expressed as follows:

$$F = x \cup y, \tag{A.1}$$

where "$\cup$" is the union operation. F equals 1 if x or y equals 1 and equals 0 if x and y both equal 0.

A.2.2 Intersection Operation

Intersection operation is usually called the logic multiply. The intersection operation of two Boolean variables x and y can be expressed as follows:

J. Li and W. Liu, *Lifeline Engineering Systems*,
https://doi.org/10.1007/978-981-15-9101-3

$$F = x \cdot y, \tag{A.2}$$

where "·" is the intersection operation. F equals 1 if x and y both equal 1 and equals 0 if x or y equals 0. Usually, "·" can be omitted.

A.2.3 *Inverse Operation*

Inverse operation is usually called the logic not. The inverse operation of a Boolean variables x is expressed as follows:

$$F = \bar{x}, \tag{A.3}$$

where "-" is the inverse operation. F equals 1 if x equals 0 and 0 if x equals 1.

The transfer relationships between the above basic logic operations are

$$x \cup y = x + y - xy \tag{A.4}$$

$$\bar{x} = 1 - x \tag{A.5}$$

$$\overline{x \cup y} = (1 - x)(1 - y) \tag{A.6}$$

The priority of these three Boolean algebra operations is that the inverse operation is the highest, the intersection operation is second higher and union operation is the lowest.

A.3 Basic Law of Boolean Algebra

A.3.1 *Commutative Law*

$$x \cup y = y \cup x \tag{A.7}$$

$$xy = yx \tag{A.8}$$

A.3.2 *Distributive Law*

$$x \cup yz = (x \cup y)(x \cup z) \tag{A.9}$$

$$x(y \cup z) = xy \cup xz \tag{A.10}$$

A.3.3 Principal Element Law

$$0 \cup x = x \tag{A.11}$$

$$1 \cdot x = x \tag{A.12}$$

$$1 \cup x = 1 \tag{A.13}$$

$$0 \cdot x = 0 \tag{A.14}$$

A.3.4 Complementary Element Law

$$x \cup \bar{x} = 1 \tag{A.15}$$

$$x \cdot \bar{x} = 0 \tag{A.16}$$

The above four laws construct the basic laws of Boolean algebra and the other laws can be derived from these basic laws. For example, the following laws can be derived from the basic laws.

A.3.5 Associative Law

$$x \cup (y \cup z) = (x \cup y) \cup z \tag{A.17}$$

$$x(yz) = (xy)z \tag{A.18}$$

A.3.6 Absorption Law

$$x \cup xy = x \tag{A.19}$$

$$x(x \cup y) = x \tag{A.20}$$

From equation (A.17), it can give that:

$$x_1 \cup (x_1 \cup x_2 \cup x_3 \cup \cdots \cup x_n) = x_1 \cup x_2 \cup x_3 \cup \cdots \cup x_n \tag{A.21}$$

Equation (A.21) is the normal expression of absorption law.

A.3.7 Idempotent Law

$$x \cup x = x \tag{A.22}$$

$$x \cdot x = x \tag{A.23}$$

A.3.8 Overlay Law

$$x \cup y = x + \bar{x}y = y + \bar{y}x \tag{A.24}$$

A.4 Disjoint Sum

Equation (A.24) is the basic equation of disjoint sum operation. In order to explain the concept of disjoint sum, mutually exclusive Boolean variables are introduced first.

For two Boolean variables x and y, if $xy = 0$, x and y are called mutually exclusive Boolean variables. This means $y = \bar{x}$. Otherwise, they are called compatible Boolean variables.

As x and $\bar{x}$ are mutually exclusive Boolean variables, x and $\bar{x}y$ are disjoint events mutually. By producing mutually exclusive Boolean variables, the union operation in the left of Eq. (A.24) transfers to the disjoint sum operation in the right of equation (A.24). Thus, Eq. (A.24) is usually called disjoint sum formula.

Equation (A.24) can be used to transfer a normal Boolean function to a mutually exclusive Boolean function (disjoint sum). For example, a Boolean function $F = (x \cup y) \cup z$ can be expressed as follows:

$$F = (x \cup y) \cup z = (x \cup y) + \overline{(x \cup y)}z = x + \bar{x}y + \overline{x}\overline{y}z \tag{A.25}$$

Independent Boolean variable is another variable related to mutually exclusive Boolean variables. In probabilistic space, if $p_r\{xy\} = p_r\{x\}p_r\{y\}$, x and y are called mutually independent Boolean variables. Otherwise, they are called dependent Boolean variables.

It should be pointed out that independence and mutually exclusion are two different concepts. Independence means the occurrence of x has no relationship with the occurrence of y. But mutually exclusion means y will not occur if x occurs and vice versa. Thus, mutually exclusion is nothing less than a dependent relation, which means the occurrence of x has a close relationship with the occurrence of y.

A.5 Main Theorems of Boolean Algebra

A.5.1 Theorem 1: De Morgan Theorem

$$\overline{x_1 \cup x_2 \cup \cdots \cup x_n} = \bar{x}_1 \cdot \bar{x}_2 \cdots \bar{x}_n \tag{A.26}$$

or

$$\overline{x_1 \cdot x_2 \cdots x_n} = \bar{x}_1 \cup \bar{x}_2 \cup \cdots \cup \bar{x}_n \tag{A.27}$$

Using the disjoint sum Eq. $x \bigcup y = x + \bar{x}y$, equation (A.27) can be rewritten as follows:

$$\begin{aligned} \overline{x_1 \cdot x_2 \cdots x_n} &= \bar{x}_1 \cup \bar{x}_2 \cup \cdots \cup (\bar{x}_{n-1} + x_{n-1}\bar{x}_n) \\ &= \bar{x}_1 + x_1\bar{x}_2 + \cdots + x_1x_2 \cdots \bar{x}_n \end{aligned} \tag{A.28}$$

Equation (A.28) is called the mutually exclusive De Morgan theorem, which is an important equation in narrowing the boundary of system reliability and transferring from the minimal path set to minimal cut set and vice versa.

Whether Boolean variables $x_1, x_2, \cdots, x_n$ are mutually independent or not, De Morgan theorem is always correct.

A.5.2 Theorem 2: Inclusive Exclusive Theorem

$$P_r\left(\bigcup_{k=1}^{m} x_k\right) = P_r\left(\sum_{j=1}^{m} (-1)^{j-1} \sum_{1 \le i_1 < \cdots < i_j \le m} \prod_{l=1}^{j} x_{i_l}\right) \tag{A.29}$$

When all the Boolean variables are mutually exclusive, Eq. (A.29) becomes

$$P_r\left(\bigcup_{k=1}^{m} x_k\right) = P_r\left(\sum_{k=1}^{m} x_k\right) \tag{A.30}$$

When all the Boolean variables are independent, Eq. (A.29) becomes

$$P_r\left(\bigcup_{k=1}^{m} x_k\right) = 1 - \prod_{k=1}^{m} (1 - p_r\{x_k\}) \tag{A.31}$$

The theorem converts the logic sum of Boolean variables to algebraic sum in the sense of probability.

Appendix B
Seismic Reliability Analysis of Transformer Substation

Transformer substations are the main components of power electricity systems. The function of transformer substation is to transfer voltage from one level to another level and distribute power within connection loops. Generally, transformer substations can be classified into step-up, step-down and transfer substations. Whether the substations operate well or not is closely related to the reliabilities of main connection systems of the substation. In power networks, an electric-voltage substation is a node of the networks. In order to evaluate the power networks reliability, the seismic reliabilities of the main connection systems should be given first (Li et al. 2000a, b, c).

B.1 Basic Category of Main Connection Systems

Usually, for step-up or step-down substations, the main connection systems can be classified into three categories, double-busbar connection, double-busbar connection with auxiliary busbar, $1\frac{1}{2}$ breaker busbar connection.

The double-busbar connection constructs a loop by connecting two disconnectors of a breaker to two busbars. Taking the main connection of step-up substations as an example, its main body is shown in Fig. B.1. Busbars 1 and 2 in the double-busbar system are both operational busbars. When it is overhauled, the two busbars can be overhauled in turn and, therefore, there is no impact on the power supply.

When an auxiliary busbar is added to the double-busbar connection, it becomes a double-busbar connection with auxiliary busbar. Therefore, the operational reliability and flexibility of such a system are improved. Figure B.2 shows a double-busbar connection with auxiliary busbar of step-up substation. When one busbar fails or is overhauled, the power will not be cutoff.

$1\frac{1}{2}$ breaker busbar connection is shown in Fig. B.3. Its characteristic is that a ring connection form is adopted. When one busbar is overhauled, loop does not need to be switched. Meanwhile, when any breaker is overhauled, loop connection need not to be changed. Under normal operation, two busbars and all breakers operate to form

J. Li and W. Liu, *Lifeline Engineering Systems*,
https://doi.org/10.1007/978-981-15-9101-3

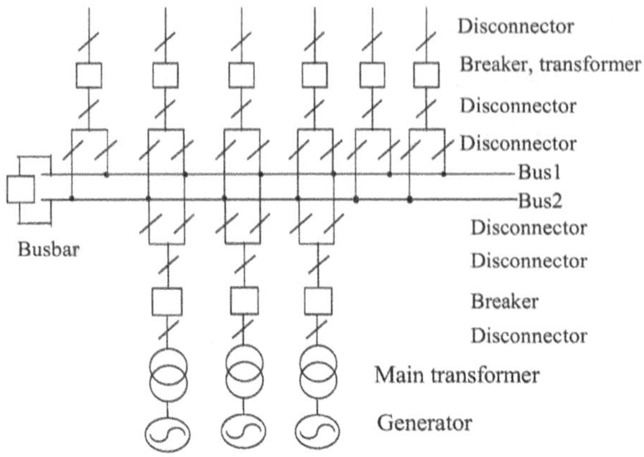

Fig. B.1 Double-busbar connection system

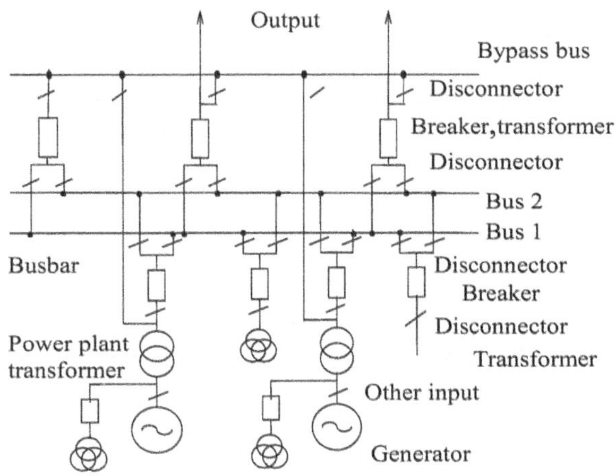

Fig. B.2 Double-busbar connection with auxiliary busbar system

a multi-loop to supply power. This connection improves system reliability greatly. Thus, it is used extensively.

B.2 Principle of Seismic Reliability Analysis of Main Connection System

The main connection system of a transformer substation is a combination of a series of high-voltage equipments, including transformers, breakers, disconnectors, current transformers and busbars. When the power sources are seen as sources and the output lines are seen as a sink, a connection network is constructed from the sources to the sinks. According to a different purpose, this network can be a single-source-single-sink network, a single-source-multi-sink network or a multi-source-multi-sink

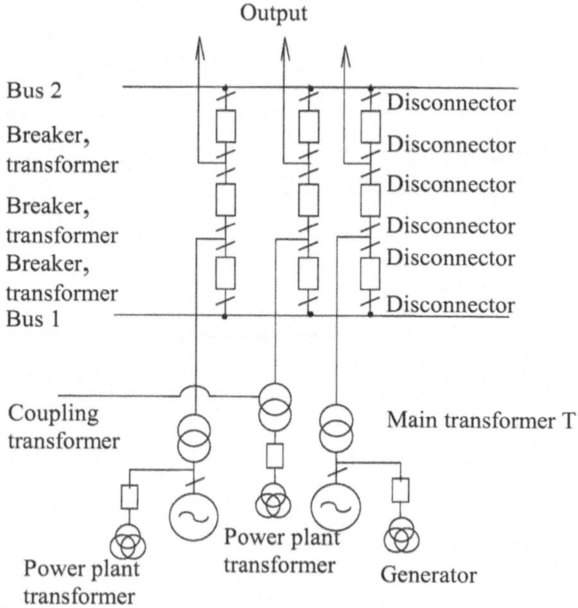

Fig. B.3 $1\frac{1}{2}$ breaker busbar connection system

network. When suffering earthquakes, the reliability of power transformation station depends on not only the reliabilities of equipments but also the main connection category, i.e. network topology. Thus, based on equipment seismic reliabilities, a system reliability analysis should be adopted to evaluate the seismic reliability of main connection system. Considering the function of substation after earthquakes, three demand levels can be established.

(1) Level 1: at least one output line can output power.
(2) Level 2: all output lines can output power.
(3) Level 3: a specified output line can output power.

Apparently, level 1 is the lowest requirement for substation, level 2 is the highest requirement and level 3 is the specified requirement considering earthquake relief. In practice, level 3 can be extended as a part of specified output lines can output power.

B.3 Seismic Reliability Analysis Model of the Main Connection System

In principle, the network connectivity analysis method illustrated in Chapter 6 can be used to analyze the seismic reliability of main connection system. In order to use the method, two assumptions should be introduced:

(1) Equipments cannot be repaired during earthquakes.

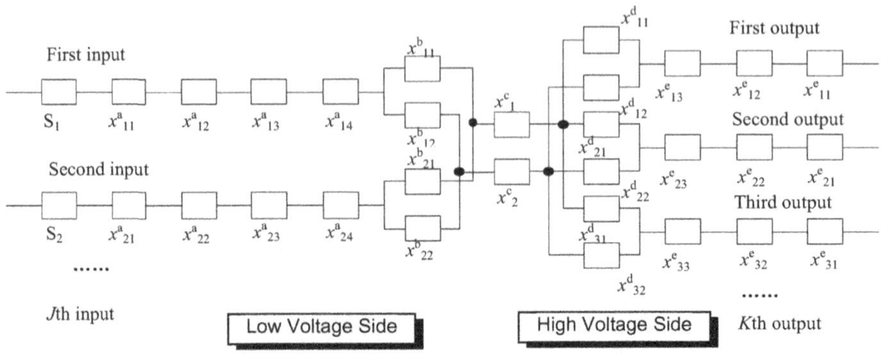

Fig. B.4 Logical block diagram of double-busbar connection of step-up substation

(2) Failure events of the equipments are independent of each other.

Assumption 1 is to distinguish it from reliability analysis of common systems considering operation and maintenance, and assumption 2 is a basic premise of classical system reliability analysis.

The process of seismic reliability analysis of main connection system can be stated as follows:

(1) Simplify the system as a logical block diagram. For simplification, some electrical elements can be transformed into equivalent elements under the condition of not changing the system logical structure.
(2) Construct system structural function according to the logical block diagram.
(3) Using system reliability analysis method, combining with the equipment reliability, to give system seismic reliability.

As an example, a double-busbar system of step-up substation is used to illustrate the process. The double-busbar system of step-up substation is shown in Fig. B.1. Its logical block diagram is shown in Fig. B.4. In this diagram, S is the input end, X^a_{ij} represents the state of the jth element in the ith line at the end of low voltage, such as transformer, breaker, disconnector, current transformer, etc.X^b_{ij} is the state of the jth disconnector in the ith line at the end of low voltage, X^c_i is the state of the ith busbar, X^d_{ij} is the state of the jth disconnect in the ith line at the end of high voltage and X^{e}_{ij} represents the state of the jth element in the ith line at the end of high voltage, such as an oil-minimum breaker, disconnector and current transformer, etc.

According to this logical block diagram, the system structural functions under different demand levels can be established as follows:

(1) **Level 1**

The structural function can be written as follows:

$$\varphi(x) = \sum_{i=1}^{2} x^c_i \left(\sum_{j=1}^{J} x^b_{ji} \prod_{m=1}^{M} x^a_{jm} \right) \left(\sum_{k=1}^{K} x^d_{ki} \prod_{n=1}^{N} x^e_{kn} \right) \tag{B.1}$$

where 2 means there are two busbars, J is the number of input loops, M is the number of diconnectors which are not connected with busbar in the input loop, K is the number of output loops and N is the number of diconnectors which are not connected with busbar in the output loop.

When x is the element seismic reliability, the seismic reliability of double-busbar connection system, according to Level 1 described in B.2, can be expressed as follows:

$$R_s = 1 - \prod_{i=1}^{2} \left\{ 1 - x_i^c \left[1 - \prod_{j=1}^{J} \left(1 - x_{ji}^b \prod_{m=1}^{M} x_{jm}^a \right) \right] \left[1 - \prod_{k=1}^{K} \left(1 - x_{ki}^d \prod_{n=1}^{N} x_{kn}^e \right) \right] \right\} \quad \text{(B.2)}$$

(2) **Level 2**

According to Level 2 described in B.2, all output line should output power. However, for input line, it will work only if power is input. Thus, its structural function is

$$\varphi(x) = \sum_{i=1}^{2} x_i^c \left(\sum_{j=1}^{J} x_{ji}^b \prod_{m=1}^{M} x_{jm}^a \right) \prod_{k=1}^{K} \left(x_{ki}^d \prod_{n=1}^{N} x_{kn}^e \right) \quad \text{(B.3)}$$

When x is the element seismic reliability, the seismic reliability of double-busbar connection system considering Level 2 can be expressed as follows:

$$R_s = 1 - \prod_{i=1}^{2} \left\{ 1 - x_i^c \left[1 - \prod_{j=1}^{J} \left(1 - x_{ji}^b \prod_{m=1}^{M} x_{jm}^a \right) \right] \prod_{k=1}^{K} \left(1 - x_{ki}^d \prod_{n=1}^{N} x_{kn}^e \right) \right\} \quad \text{(B.4)}$$

(3) **Level 3**

According to Level 3 of the analysis, only a specified output line should be considered. Thus, its structural function is

$$\varphi(x) = \sum_{i=1}^{2} x_i^c \left(\sum_{j=1}^{J} x_{ji}^b \prod_{m=1}^{M} x_{jm}^a \right) \prod_{n=1}^{N} x_{Tn}^e \quad \text{(B.5)}$$

where T is the specified output line.

Similarly, the seismic reliability of double-busbar connection system considering level 3 can be expressed as follows:

$$R_s = 1 - \prod_{i=1}^{2} \left\{ 1 - x_i^c \left[1 - \prod_{j=1}^{J} \left(1 - x_{ji}^b \prod_{m=1}^{M} x_{jm}^a \right) \right] \prod_{n=1}^{N} x_{Tn}^e \right\} \quad \text{(B.6)}$$

When the seismic reliabilities of high-voltage equipments in the system are calculated using the method in chapter 5, the system reliability under different level can be given easily. In literature (Li et al. 2000a, b, c), the seismic reliability of various main connection system has been investigated in detail.

Appendix C
Seismic Secondary Fire Analysis

C.1 Seismic Secondary Fire Model

When a city is divided into a number of districts, Poisson random field model can be used to simulate the occurrence of seismic secondary fire (Li and Li 1992, Li and Jiang 2000). According to the model, the probability that K fires occur in a district in T days after earthquakes is

$$P(K) = \frac{\lambda^K e^{-\lambda}}{K!} \tag{C.1}$$

where λ is the fire occurrence rate, and it can be given by (Li and Jiang 2000):

$$\lambda = \frac{\rho}{\rho_c}\mu_m A_m + \rho A \tag{C.2}$$

where ρ is the fire occurrence rate without earthquakes in the district, ρ_c is the fire occurrence rate without earthquakes in the city, A_m is the area of buildings which suffered medium or more serious damages under earthquakes in the district, μ_m is the fire density of A_m and generally takes the value of 0.03~0.06/10^4 m^2, A is the total building area in the district.

As the time interval between two adjacent secondary fires follows an exponential distribution with parameter λ, the following numerical simulation process can be used to simulate the occurrence of seismic secondary fire in the district.

(1) Produce random numbers $(x_1, x_2, \ldots)$ which follow uniform distribution between 0 and 1.
(2) Use the following inverse transformation to produce seismic secondary time interval:

J. Li and W. Liu, *Lifeline Engineering Systems*,
https://doi.org/10.1007/978-981-15-9101-3

$$\Delta t_i = -\frac{1}{\lambda} ln x_i \tag{C.3}$$

(3) Use the time interval Δt_i to simulate the occurrence of secondary fire in the district.

C.2 Statistical Parameters of Firefighting

For fighting seismic secondary fire, the firefighting process can be simplified into four main stages: (1) from fire occurrence to receiving alarm, (2) from receiving alarm to starting off, (3) from starting off to arriving at fire location, (4) from arriving at fire location to putting off the fire. According to the data in the daily firefighting process in China, the main statistical parameters can be given as follows:

(1) The time from fire occurrence to receiving alarm t_1

Usually, t_1 follows gamma distribution and its probability density function is

$$f(t_1) = \frac{-3.62^{-1.1} t_1^{0.1} e^{-\frac{t_1}{3.62}}}{\Gamma(1.1)} \tag{C.4}$$

where $\Gamma(\cdot)$ is the gamma function.

(2) The time from receiving alarm to starting off t_2

According to the statistics of 2851 fire case, $t_2 = 1$ minute

(3) The time from starting off to arriving at the fire location t_3

According to the damage state of roads after earthquakes, the minimal path between firefighting team and fire location can be identified. During the simulation of firefighting process, the average velocity of truck may be set as 20–30 km/h and then t_3 can be calculated.

(4) The time from arriving at fire location to putting off the fire t_4

Usually, t_4 can be given by using the following statistical distribution:

$$f(t_4) = \frac{-6.81^{-2.38} t_4^{1.38} e^{-\frac{t_4}{6.81}}}{\Gamma(2.38)} \tag{C.5}$$

Obviously, the probability density function can be given statistically for the specified city.

Bibliography

Ahmad, S. H., & Jamil, A. T. M. (1982). A modified technique for computing network reliability. *IEEE Transactions on Reliability, R-31*(1), 41–44.

Ai, X. Q., & Li, J. (2005). Research on seismic response of underground pipelines in solid-liquid media. *Earthquake Engineering and Engineering Vibration, 25*(2), 136–140. (in Chinese).

ATC. (2010). Tentative provisions for the development of seismic regulations for buildings, ATC-3-16.

Ayala, A. G., & O'Rourke, M. J. (1989). Effects of the 1985 Michoacan earthquake on water systems and other buried lifelines in Mexico. *National Center for Earthquake Engineering Research.*

Bao, Y. F., Li, J., Liu, W., et al. (2006). Aseismic reliability analysis of shenyang city's natural gas system. *Earthquake Resistant Engineering and Retrofitting, 28*(5), 107–110. (in Chinese).

Benuska, L. (1990). Loma Prieta earthquake reconnaissance report. *Earthquake Engineering Research Institute.*

Cao, J., & Cheng, K. (1986). *Reliability mathematics.* Beijing: Science Press. (in Chinese).

Chen, L. L. (2002). Reliability analysis and optimization of aseismic function of urban water supply network system (Ph.D. Dissertation). Tongji University, China. (in Chinese).

Chen, L. L., & Li, J. (2004). Aseismatic serviceability analysis of water supply network. *Engineering Mechanics, 21*(4), 45–50. (in Chinese).

Chen, J. B., Ghanem, R., & Li, J. (2009). Partition of the probability space in probability density evolution analysis of non-linear stochastic structures. *Probabilistic Engineering Mechanics, 24*(1), 27–42.

Chen, J. B., & Li, J. (2013). Optimal determination of frequencies in the spectral representation of stochastic processes. *Computational Mechanics, 51*(5), 791–806.

Codification Committee for Wenchuan Earthquake Disaster. (2009). *Wenchuan earthquake disaster.* Chengdu: Chengdu Cartographic Press. (in Chinese).

Cornell, C. A., & Vanmarcke, E. H. (1969). The major influences on seismic risk. In *Proceedings of the 4th World Conference of Earthquake Engineering*, vol. 1, pp. 69–83.

Der Kiureghian, A., & Hofer, A. (1992). Response spectrum method for multi-support seismic excitations. *Earthquake Engineering and Structural Dynamics, 21*(8), 713–740.

Dorigo, M., Maniezzo, V., & Colorni, A. (1996). Ant system: Optimization by a colony of cooperating agents. *IEEE Transactions on Systems, Man, and Cybernetics, Part B: Cybernetics, 26*(1), 1–13.

Hada, Y., & Meguro, K. (2000). Optimum restoration model considering interactions among lifeline systems-interactions among restoration activities of lifeline utilities. In *12th World Conference on Earthquake Engineering.* Auckland, New Zealand: ASCE: 8 p (CD-ROM).

Hall, J. F. (1996). Northridge earthquake of January 17, 1994 reconnaissance report. *Earthquake Engineering Research Institute.*

J. Li and W. Liu, *Lifeline Engineering Systems*,
https://doi.org/10.1007/978-981-15-9101-3

Han, X., Li, J., & Shen, Z. Y. (1998). Petri net based methodology for dispatching and commanding progress of urban fire-fighting. *Journal of Natural Disaster, 7*(3), 25–31. (in Chinese).

Heredia-Zavoni, E., & Vanmarcke, E. H. (1994). Seismic random-vibration analysis of multisupport-structural systems. *Journal of Engineering Mechanics, 120*(5), 1107–1128.

Hou, Z. L., Hu, Y. X. Takada, S. & Wang, L. R. L. (1990). In *Proceedings of the China-Japan-USA Trilateral Symposium on Lifeline Earthquake Engineering*. Beijing, China.

Hu, Y. X. (1992). *Comprehensive probability method for seismic hazard analysis*. Seismological Press. (in Chinese).

Hu, Y. X., Takada, S., & Wang, L. R. L. (1994). In *Proceedings of the China-Japan-USA Trilateral Symposium on Lifeline Earthquake Engineering*. Xian, China.

Hwang, H. H., Lin, H., & Shinozuka, M. (1998). Seismic performance assessment of water delivery systems. *Journal of Infrastructure Systems, 4*(3), 118–125.

Investigation group of Japan Osaka Great Earthquake. (1997). *Investigation Report on the Great Osaka Earthquake in Japan*. Seismological Press. (in Chinese).

Jiang, J. H. (2001). GIS-based seismic performance simulation analysis of composite lifeline system (Ph.D. Dissertation). Tongji University, China. (in Chinese).

Kiureghian, A. D., & Ang, A. H. S. (1977). A fault-rupture model for seismic risk analysis. *Bulletin of the Seismological Society of America, 67*(4), 1173–1194.

Lee, K. H., & Jung, M. Y. (1994). Petri net application in flexible process planning. *Computers Industrial Engineering, 27*(4), 505.

Li, J. (1997). Disaster response analysis and system control of composite engineering system. *Journal of Natural Disaster, 6*(3), 1–9. (in Chinese).

Li, J. (1998). Earthquake response analysis and system control of lifeline composite system. In *Proceedings of 3rd China-Japan-US trilateral Symposium on Lifeline Earthquake Engineering* (pp. 231–239). China: Kunming.

Li, J. (2002). Seismic reliability analysis for lifelines. In *Proceedings of 4th China-Japan-US trilateral symposium on lifeline earthquake engineering* (pp. 283–290). China: Qingdao.

Li, J. (2005). *Aseismic theory and application of lifeline engineering systems*. Science Press of China. (in Chinese).

Li, J. (2007a). Progress in lifeline engineering researches. *Frontiers of Civil Engineering in China, 1*(1), 34–40.

Li, J. (2007b). Seismic reliability analysis and topology optimization of lifeline networks, (Keynote Lecture). In *Proceedings of 5th China-Japan-US Trilateral Symposium on Lifeline Earthquake Engineering* (pp. 23–34). China: Haikuo.

Li, J. (2013). Reliability-based design of lifeline systems: Recent developments and possible tendency in the future, (Keynote Lecture). In *Proceedings of 6th China-Japan-US Trilateral Symposium on Lifeline Earthquake Engineering*. China: Chengdu.

Li, J. (2016). Probability density evolution method: Background, significance and recent developments. *Probabilistic Engineering Mechanics, 44*, 111–117.

Li, J. (2018a). *Urban earthquake disaster field control*. Tongji Univesity Press.

Li, J., He, J., & Li, T. (2002). Seismic reliability analysis of large electric power network. *Journal of Haerbing Jianzhu University, 35*(1), 7–11. (in Chinese).

Li, J., & Li, G. Q. (1992a). *Introduction to earthquake engineering*. Seismological Press. (in Chinese).

Li, J., & Li, G. Q. (1992b). A research on the prediction model of earthquake-arouse fire disaster. *Earthquake Research in China, 8*(1), 76–82. (in Chinese).

Li, J., Liu, X. T., & Liu, W. (2007). Application of ant colony algorithm to seismic topological optimization of lifeline network systems. *Journal of Disaster Prevention and Mitigation Engineering, 27*(2), 127–132. (in Chinese).

Li, J., & Song, J. X. (1995). Study on the simulation of post-earthquake fire. *Sino-US symposium on post-earthquake rehabilitation and reconstruction* (pp. 100–107). China: Kunming.

Li, J., & Song, J. X. (1996c). Study on the system simulation and simulation-based control for earthquake disaster field. In *Proceedings of 11th World Conference on Earthquake Engineering*. Mexico: Acopulco.

Li, J., & Song, J. X. (1996a). Study on simulation-based control for earthquake disaster field in Kaifeng city. *World Earthquake Engineering, 3,* 20–26. (in Chinese).

Li, J., Wei, S. L., & Liu, W. (2006). Seismic reliability analysis of urban water distribution network. *Earthquake Engineering & Engineering Vibration, 5*(1), 71–78.

Li, J., Wei, S. L., & Liu, W. (2009). Seismic reliability optimization of water distribution networks with simulated annealing algorithms. *Earthquake Engineering and Engineering Vibration, 29*(3), 108–114. (in Chinese).

Li, J., & Xing, Y. (2010). Reliability-based seismic design of lifeline networks. *Journal of Tongji University (Nature Science), 38*(6), 783–786. (in Chinese).

Liu, W., & Li, J. (2008). Stochastic seismic response of pipelines with corrosion. *Journal of Earthquake Engineering, 12*(6), 914–931.

Liu, W., & Li, J. (2009). An improved recursive decomposition algorithm for reliability evaluation of lifeline networks. *Earthquake Engineering & Engineering Vibration, 8*(3), 409–420.

Liu, W., & Li, J. (2012). An improved cut-based recursive decomposition algorithm for reliability analysis of networks. *Earthquake Engineering & Engineering Vibration, 11*(1), 1–10.

Liu, W., & Li, J. (2014). Comparison of algorithms for seismic topology optimization of lifeline networks. *Structure and Infrastructure Engineering, 10*(11), 1357–1368.

Liu, W., Li, Z. C. L., Song, Z. Y., & Li, J. (2018). Seismic reliability evaluation of gas supply networks based on the probability density evolution method. *Structural Safety, 70,* 21–34.

Liu, W., Miao, H. Q., Wang, C., & Li, J. (2017a). Experimental validation of a model for seismic simulation and interaction analysis of buried pipe networks. *Soil Dynamics and Earthquake Engineering, 100,* 113–130.

Liu, W., Sun, Q. W., Miao, H. Q., & Li, J. (2015a). Nonlinear stochastic seismic analysis of buried pipeline systems. *Soil Dynamics and Earthquake Engineering, 2015*(74), 69–78.

Liu, W., Xu, L., & Li, J. (2017b). Seismic design of water distribution networks using auto-generation techniques. *Structure and Infrastructure Engineering, 13*(7), 831–843.

Liu, W., Zhao, Y. G., & Li, J. (2015b). Seismic functional reliability analysis of water distribution networks. *Structure and Infrastructure Engineering, 11*(3), 363–375.

Loh, C. H. (1985). Analysis of the spatial variation of seismic waves and ground movements from SMART-1 array data. *Earthquake Engineering and Structural Dynamics, 14,* 561–581.

Miao, H. Q., Liu, W., Wang, C., & Li, J. (2016). Artificial earthquake test of gas supply networks. *Soil Dynamics and Earthquake Engineering, 90,* 510–520.

Nojima, N., & Kameda, H. (1996). Lifeline interaction in the Hanshi-Awaji earthquake disaster. In *The 1995 Hyogoken-Nanbu Earthquake-Investigation into Damage to Civil Engineering Structures*. Japan Society of Civil Engineers.

O'Rourke, T. D. (1996). Lessons learned for lifeline engineering from major urban earthquakes. In *11th World Conference of Earthquake Engineering*. Acapulco, Mexico: ASCE.

Oliveira, C. S., Hao, H., & Penzien, J. (1991). Ground motion modeling for multiple-input structural analysis. *Structural Safety, 10*(1–3), 79–93.

Patwardhan, A. S., Kulkarni, B. R., & Tocher, D. (1980). A semi-Markov model for characterizing recurrence of great earthquakes. *Bulletin of the Seismological Society of America, 70*(1), 323–347.

Shao, Z.Y. (2001). Piecewise linear optimization model for water supply Network (M.S dissertation). Tongji University, China. (in Chinese).

Shinozuka, M., & Tanaka, S. (1996). Effects of lifeline interaction under seismic conditions. In *11th World Conference on Earthquake Engineering*.

State Key Laboratory of Disaster Reduction in Civil Engineering. (2008). *Earthquake disaster of wenchuan earthquake*. Shanghai: Tongji University Press. (in Chinese).

Sun, Q. W., Liu, W., & Li, J. (2011). Analysis model of buried pipeline network under earthquake ground motions. *Journal of Disaster Prevention and Mitigation Engineering, 31*(6), 654–658. (in Chinese).

Trifunac, M. D., & Lee, V. W. (1990). Frequency dependent attenuation of strong earthquake ground motion. *Soil Dynamics and Earthquake Engineering, 9*(1), 3–15.
Tsai, J. P., Yang, S. J., & Chang, Y. H. (1995). Timing constraint Petri nets and their application to schedulability analysis of real-time system specifications. *IEEE Transactions on Software Engineering, 21*(1), 32–49.
Wang, C., Liu, W., & Li, J. (2015). Artificial earthquake test of buried water distribution network. *Soil Dynamics and Earthquake Engineering, 79,* 171–185.
Wang, D. R. (1983). *Nonlinear equation method and optimization method (in Chinese).* People's Education Press.
Wang, F. (1986). A renewal process model of earthquake occurrence for seismic hazard estimation. *Earthquake Engineering and Engineering Vibration, 6*(2), 17–26.
Wei, S. L., Li, J., & Zhou, J. H. (2008). A seismic reliability analysis of Shenyang city water supply system. *World Earthquake Engineering, 24*(4), 106–110. (in Chinese).
Xing, Y., & Li, J. (2008). Hybrid optimization strategy for topology design of lifeline networks. *Journal of Tongji University (Nature Science), 36*(5), 569–573. (in Chinese).
Xiong, G. L., Xiao, T. Y., Zhang, Y. Y. (1991). *Simulation of continuous system and discrete event system.* Tsinghua University Press. (in Chinese).
Xuan, G. N., & Ceng, R. W. et al. (2004). *Genetic algorithm and engineering optimization.* Tsinghua University Press. (in Chinese).
Yu, H., et al. (2000). A combined genetic algorithm/simulated annealing algorithm for large scale system energy integration. *Compputer and Chemical Engineering, 24,* 2023–2035.
Zhou, X. Y., Su, J. Y., & Wang, G. J. (1985). Seismic hazard analysis of the basis of Neyman-scott model. *Earthquake Engineering & Engineering Vibration, 5*(2), 17–26.
Zhou, X. Y., & Takada, S. (1998). In *Proceedings of the China-Japan-USA Trilateral Symposium on Lifeline Earthquake Engineering*. China: Kunming.
Zhou, X. Y., & Takada, S. (2002). In *Proceedings of the China-Japan-USA Trilateral Symposium on Lifeline Earthquake Engineering*. China: Qingdao.

References

Li, J., & Li, G.Q. (1992). A research on the prediction model of earthquake-arouse fire disaster. *Earthquake research in China, 8*(1): 76–82. (in Chinese).
Li, J., & Jiang, J. H. (2000). Hazard nalysis of urban post-earthquake fire. *Journal of Natural Disaster, 9*(2), 87–92. (in Chinese).
Li, T., Li, J., Shen, Z. Y. (2000a). Seismic reliability analysis of breaker-and-a-half configuration of high voltage substations. *World Information of Earthquake Engineering, 16*(3), 58–62. (in Chinese).
Li, T., Li, J., Shen, Z. Y. (2000b). Seismic reliability of high voltage substation (Part 1). *Earthquake Engineering and Engineering Vibration, 20*(3), 43–49. (in Chinese).
Li, T., Li, J., Shen, Z. Y. (2000c). Seismic reliability of high voltage substation (Part 2). *Earthquake Engineering and Engineering Vibration, 20*(4), 37–43. (in Chinese).

The manufacturer's authorised representative in the EU is Springer Nature Customer Service Centre GmbH, Europaplatz 3, 69115 Heidelberg, Germany. If you have any concerns regarding our products, please contact ProductSafety@springernature.com

Printed and bound by CPI Group (UK) Ltd, Croydon, CR0 4YY
30/09/2026
02225292-0001